Dom Schopf
May 79

D0983841

Fossils
and
Progress

TABLE OF STRATA AND ORDER OF APPEARANCE OF ANIMAL LIFE UPON THE EARTH.

Table illustrating the order of geological strata and the development of animal life. From Richard Owen's *Palaeontology* (1860).

Fossils and Progress

Paleontology and the Idea of Progressive Evolution
in the Nineteenth Century

PETER J. BOWLER

Science History Publications
New York   1976

Science History Publications
a division of
Neale Watson Academic Publications, Inc.
156 Fifth Avenue, New York, N.Y. 10010

© Science History Publications 1976

Library of Congress Cataloging in Publication Data
Bowler, Peter J.
   Fossils and progress.
   Bibliography: p.
   Includes index.
   1. Paleontology—Europe—History. 2. Evolution.
I. Title.
QE705.E8B68   560'.9'034   75-40005
ISBN 0-88202-043-9

Designed and manufactured in the U.S.A.

Contents

Preface

The extent to which pre-Darwinian geology and natural history were influenced by religious and philosophical issues has been common knowledge among historians of science at least since the publication of Professor Charles Gillispie's *Genesis and Geology*. There have been numerous recent accounts of the history of evolutionism which have touched on the same theme. Yet for some time surprisingly little attention was paid to the technical advances in geology and paleontology which formed the background to these mid-nineteenth century debates. There have been few attempts to expand upon the pioneering work of Frank D. Adams, Sir Archibald Geikie, and Karl von Zittel—the emphasis has shifted almost completely to the more philosophical issues involved. To some extent this has been a beneficial movement, since greater depth of interpretation was certainly needed. But the pendulum seems, as usual, to have swung too far. In some ways the concentration on broader issues has obscured the underlying background of expanding geological and paleontological knowledge, with the result that historians have not always been aware of the true nature of the ammunition used by both sides in the great debates. The paleontologists of the nineteenth century produced the first roughly accurate outline of the history of life on the earth. Many of them were opponents of evolution, but they had to adapt their feelings on this issue to a rapid and sometimes irregular growth of the fossil evidence. Their efforts are thus of vital importance to the history of evolutionism, despite their opposition to "evolution" itself.

Only recently have the technical developments been more thoroughly explored within the deeper context provided by analysis of the philosophical issues, in, for instance, Professor M. J. S. Rudwick's *The Meaning of Fossils* and Professor Leonard G. Wilson's work on Lyell. Without such efforts, the present work would have had to be much longer in order to provide the necessary background material. I have not tried to give the same depth of technical analysis here, since my own interests derive more from intellectual history. But in developing this account of a particularly crucial concept relating to the rise of evolutionism—that of "progression"—I have tried to provide a synthesis of both the technical and the wider issues, revealing the full depth of the debates while giving some impression of the complex practical problems with which the naturalists grappled.

The research for this project was conducted while studying and teaching at the Institute for the History and Philosophy of Science and Technology at the University of Toronto. I should like to thank Professors William E. Swinton and Mary P. Winsor for their advice at this stage, and the University of Toronto Library for help in locating original sources. The actual writing was done while teaching at the Universiti Sains Malaysia in Penang; I am grateful to Professor John C. Greene for his encouragement during this period.

Peter J. Bowler

University of Winnipeg
September, 1975

1

The Origin and Structure
of the Idea of Progression

In the course of the Renaissance and the seventeenth century, European thinkers began increasingly to puzzle over the nature and significance of fossils.[1] It was not until the Enlightenment of the eighteenth century, however, that a significant number of naturalists and philosophers came to accept that these petrified figures really were the remains of once living animals and plants. A whole new perspective began to open up, in which the traditional Biblical story of the creation was replaced by a general feeling that the earth is ancient and has undergone many changes in the course of its history. Since it was not always possible to relate them to known creatures, fossils served to demonstrate that the earth's population has changed along with its physical structure. The more adventurous Enlightenment thinkers began to speculate on the means by which these changes were brought about, some of them arriving at theories which bear at least a superficial resemblance to modern evolutionism.[2] Even when the Enlightenment collapsed and was replaced by the far more conservative outlook of the early nineteenth century, the idea of a long period of development was retained and was at first adapted to the outlines of Genesis. The enormous expansion of geology and paleontology at this time confirmed that there has been a sequence of different populations in the course of the earth's history, and an outline of the process of development began to emerge. This outline served as a background against which the evolutionary debates of the mid-nineteenth century were fought out, with fossil evidence being used on both sides and helping to mold the opinions of both the opponents and the supporters of evolution.

One of the most important inferences to be derived from the fossil record was the apparent progressive development of life, in the course of the earth's history, from the simplest to the most complex forms. The possibility of such a progression had already been hinted at during the Enlightenment, but only in the following century did it gradually acquire a widespread popularity. The idea would presumably have been generated in any case by the fossil evidence alone, but there can hardly be a coincidence in the fact that its rise to popularity shows a close parallel with the growth of the belief in social progress.[3] The optimistic social philosophers of the Victorian era must have seen the progressive development of life as a perfect corollary to their own creed of progress, and the two effects came more and more to be associated as man was integrated into nature through a theory of evolution. But the concept of biological progress was no more simple or straightforward than its social counterpart; indeed, the various possible interpretations in both fields show a certain amount of parallelism. Efforts to relate the different kinds of progressionism to the ever expanding fossil evidence were among the most significant technical factors influencing the rise of modern evolutionism. But it is impossible to study these efforts without at the same time being aware of their intimate relationship to a number of vital human problems.

Some of the Enlightenment naturalists had already tackled the question of whether or not the history of life exhibits a definite pattern of development. Certain theories of the earth held that drastic physical changes have affected the planet, hence implying that there must have been a corresponding sequence of changes in the nature of its inhabitants. The comte de Buffon's theory, for instance, postulated that the earth has gradually cooled down from its originally molten state, and that the earlier forms of life were necessarily adapted to higher temperatures.[4] But the possibility was also raised that the history of life has a logic or pattern of its own, to some extent independent of the changing conditions. In particular the idea of the progressive development of life arose out of what A.O. Lovejoy called the "temporalization" of the ancient chain-of-being concept.[5] The chain represented an absolute, linear hierarchy of organic forms stretching from the simplest up to man himself. Some thinkers, especially Charles Bonnet and J.B. Robinet, held that the chain could be regarded as the plan of organic development, defining the stages in a divinely planned hierarchy along which life has progressed in the course of the earth's history. In his *Palingénésie philosophique* of 1769, Bonnet argued that:

> If God's plan required that the sentient beings inhabiting a certain planet should pass successively through various subordinate degrees of perfection, He would pre-establish from the beginning the means destined to increase the form of their perfection and to give them the whole extension that their nature can bear.[6]

The belief that life has necessarily advanced along a scale of organization was also incorporated into the last and most sophisticated of the Enlightenment's evolutionary theories, that of the chevalier de Lamarck. Taking up a common theme of eighteenth century materialism, Lamarck argued that life could be spontaneously generated from inert matter by the activity of a vital fluid. But unlike some of his predecessors, he insisted that only the simplest organisms could be formed in this way and that these would then form the starting point for a long and gradual ascent of the scale toward higher forms.[7] Whether or not this belief was derived from the traditional concept of the chain-of-being is still a subject of debate—as is the whole of Lamarck's theory—but the ascent of a scale of organization was certainly central to his system.[8] Lamarck also accepted that the ascent would be modified by the need of the organisms to adapt to gradually changing physical conditions, but this was not the cause of the progression, since his geological theory was uncompromisingly uniformitarian.[9]

These earlier ideas of progression had been developed largely as speculations—for all his work on invertebrate fossils, even Lamarck does not seem to have felt that it would be possible to use paleontological evidence to demonstrate that there has actually been an advance of life as predicted by his theory. But the development of vertebrate paleontology by Georges Cuvier and his followers in the early decades of the new century began at last to provide definite evidence that the history of life has followed some kind of progressive pattern. As it first became available, however, the fossil record was certainly not enough to support any particular explanation of how and why the progression occured, and in particular it did not appear to confirm Lamarck's belief that it was a continuous process. Many naturalists developed theories of progression based on successive miraculous creations, while the few who did want to argue for transmutation were forced to grapple with the problem of reconciling this theory with a markedly discontinuous fossil record. In pre-Darwinian evolutionary debates such as the one centered on Robert Chambers' anonymously published *Vestiges of the Natural History of Creation* (1844), the question of the degree of continuity within the fossil progression was crucial. Everyone was concerned about progression, because it appeared to be the key which linked both taxonomy and geology with the wider issues that concerned every naturalist, issues raised by the growing need to undertake a reformation of the traditional views on the relationship between God, man, and nature. Only with the completion of this revolution through the introduction of Darwin's theory did the emphasis finally begin to shift away from progression, since natural selection not only promoted a revision of what the concept could mean but also de-emphasized its general significance in the history of life. This partial rejection of progression could almost be treated as a symbol of Darwin's advance beyond the earlier views of the development of life. The rise and fall of the progressionist viewpoint is thus a key factor in the process by which the modern theory of evolution emerged in the course of the nineteenth century.

Any theory of progression has two basic components, the first being a hierarchy or scale of organization which defines what is meant by an advance toward "higher" forms, the second a time scale against which this advance is plotted. The latter of these requirements proved by far the easier to establish. Drawing together the work of figures such as Cuvier and William Smith, the paleontologists and stratigraphists of the early nineteenth century began to map out a sequence of geological formations which represented the history of the earth's surface. The sequence of deposits they established was held to correspond to a series of time periods valid for all geographical areas. Beneath the most recent materials lay the rocks deposited during the Tertiary era, containing numerous remains of mammals. By 1811, Cuvier and Alexandre Brongniart had already divided the Tertiary into a number of sub-periods, the most important of which were later given the names Eocene, Miocene, and Pliocene by Charles Lyell. Beneath the Tertiary rocks lay the Secondary series, ranging from the chalk deposits (Cretaceous system) down to the New Red Sandstone rocks or Triassic system. The Secondary rocks contained the fossils of many vertebrates, but few mammals. The Transition formations which lay beneath were at first thought to contain only invertebrate remains, although fossil fishes were eventually discovered in great numbers, especially in the Old Red Sandstone or Devonian system. The whole sequence rested on the azoic Primary rocks. By the mid-nineteenth century the series of formations was well established, although absolute dating was impossible. A few changes have been made since the series was first defined, and the diagram given here illustrates both the original and the modern system.

Since the sequence of geological periods was generally accepted, the most basic debate over progression centered on whether or not the series of fossils did provide good evidence for an advance of life toward higher forms. In the first half of the century it was still possible to deny this altogether; Charles Lyell advanced such a position in his *Principles of Geology* of 1830 and continued to defend it for the next thirty years. As writers such as R. Hooykaas and M.J.S. Rudwick have pointed

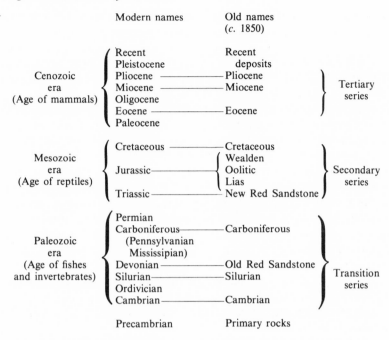

The sequence of geological periods as recognized today and in the mid-nineteenth century (note that some of the modern names were coming into use by 1850, so that the Old Red Sandstone, for instance, was already being referred to as part of the Devonian system).

out, Lyell's uniformitarian theory of the earth was distinguished from the opposing "catastrophism" as much by its emphasis on a steady-state view of the earth's history as by its rejection of catastrophes.[10] He held that geological activity was no more intense in the past because the planet's history has not been a directional process (as, for instance, with Buffon's cooling-earth theory) but an indefinite cycle of minor fluctuations. Driven by this uniformitarian philosophy, Lyell decided that there could be no absolute direction in the history of life, and argued that the evidence was by no means sound enough to prove that there were not at least small numbers of all classes of animals and plants, high and low, living in even the earlier geological periods. Enough anomalies in the record were discovered during the first half of the century to make this by no means a completely implausible opinion. Although Lyell's tenacious support was inspired largely by his overall cosmology, a small number of other naturalists joined him in rejecting progression altogether. Their arguments and discoveries served as a constant goad urging their opponents to define exactly what the basis of progressionism ought to be.

Lyell did not, however, challenge one of the progressionists' most basic assumptions. Although he felt that the fossil evidence for their belief was unsound, he had no doubt as to what would have constituted an advance of life toward higher forms. Like most of his contemporaries he accepted the notion of a hierarchy of organization, especially with respect to the vertebrates. It was generally held that the fishes were the simplest or most primitive class of the vertebrate type, followed in turn by the reptiles, birds, and mammals in ascending order of complexity. This system resembled, at least superficially, the ancient concept of the great chain of being. The degree of resemblance is debatable: it has long been a commonplace among historians of biology that Cuvier's division of the animal kingdom into four types destroyed the prospect of creating a unilinear sequence of organic forms. This thesis has been stated in greater depth by Michel Foucault, who argues that the classical natural history of the eighteenth century differed fundamentally from the kind of biology that resulted from Cuvier's reformation of taxonomy.[11] Foucault maintains that the naturalists of the Enlightenment, whatever their superficial differences of opinion, all accepted nature as a plenum of forms whose relationships could in principle (but perhaps not in practice) be expressed by an account of their physical resemblances. The chain of being was merely the simplest structure that could be imagined for the plenum. Cuvier destroyed not only the chain of being but also the whole concept of a plenum of forms: he concentrated on the inner structures of living things and expressed them as adaptive variants upon the four archetypes. Yet, although most of them professed to follow Cuvier, some of the early nineteenth century naturalists do not seem to have appreciated the full implications of this revolution. For them there was still a relatively unambiguous hierarchy of organization within the animal kingdom, most obviously among the vertebrate classes. A few even continued to imagine that the sequence had a definite limit—Louis Agassiz's claim that man represents the logical end-point of the vertebrate type's development is the best example. Everyone realized that the series of forms was not really linear, but at least some of the old attitudes remained. The conviction that there was an obvious hierarchy among the vertebrates served as the basis for measuring the supposed progression of life, once the fossil record began to give some indications of the order in which the classes appeared.

This attempt to retain a linear view of nature had begun to break down well before the *Origin of Species* appeared. To some extent this was inevitable if the full impact of Cuvier's taxonomy was to be felt. The attempt to retain a vestige of linearity was an anachronism in the nineteenth century, and the rapid expansion of the fossil collections seems to have played a significant role in convincing naturalists of this, by revealing to them the artificiality of the simple progressive scheme of living development. Real problems began to emerge as soon as the progressionists tried to define the hierarchy of organization in more detail than a mere ranking of the classes. This became a crucial issue as soon as the question was raised as to whether or not the advance of life was a continuous process. In general, transmutation theories such as that of Chambers' *Vestiges* depended upon the continuity of progression, i.e., on showing that each class reached its peak and was then succeeded by the lowest form of the next higher class. Most creationists, on the other hand, were only too glad to accept that the progression was irregular or discontinuous, since this would prove that the succession of forms could only have been introduced miraculously. To debate this issue it was necessary to define the hierarchy of organization in some detail. The hierarchy *within* each class had to be established so that it could be determined if progression occurred at this level as well as in the basic sequence of the classes themselves. In some cases there was little difficulty in deciding which were the highest and lowest orders within a class. The amphibians were generally regarded as the lowest, the most fish-like order of the reptiles. (Modern practice is to treat the amphibians and reptiles as separate classes; the great differences between them were pointed out in 1805 by Alexandre Brongniart and H.M.D. de Blainville, but many naturalists continued to treat the amphibians as merely an order of reptiles.) The marsupials were the lowest mammals, while man was clearly the highest member of this class and hence of the whole vertebrate type. In these cases the question of whether or not a progression had occurred within the class was easy to determine with respect to the fossil record available at the time. The fact that the first reptiles, for instance, did not appear to be amphibians counted heavily against the theory of continuous progression until further discoveries altered the picture in the late 1840s. But it was not always such a straightforward procedure, and the attempt to establish a hierarchy within the fishes became something of a taxonomic scandal. Opponents of transmutation such as Hugh Miller argued that the earliest known fish were the highest, since they apparently had well-developed nervous systems. Chambers, on the other hand, supported the continuity of progression by claiming that

M	HIGH	Man
A		
M		
M		
A		
L		
S	LOW	The marsupials
B	HIGH	⎫
I		⎬ This class often ignored by paleontologists because of lack of
R		fossil evidence.
D		⎬
S	LOW	⎭
R	HIGH	The dinosaurs—of all the then-known reptiles, considered to be
E		the closest to the birds and mammals.
P		
T		
I		
L		
E		
S	LOW	The amphibians—the most fish-like "reptiles."
F	HIGH	The osseous fish ⎫
I		⎬
S		⎬ This order modified and even reversed
H		by some authorities.
E		⎬
S	LOW	The cartilaginous fish ⎭

Illustrating at a superficial level the type of hierarchical arrangement of the vertebrates envisioned by some nineteenth century paleontologists.

these forms were lowly organized because they lacked a fully ossified vertebral column. The great anatomist Richard Owen actually reversed his opinion on this particular issue around 1850, and eventually was driven to admit that there was really no way of establishing an absolute hierarchy within the fishes. The same form could combine both high and low features, making the whole question of progression meaningless. As more fossils were discovered they began to fit together to show that the classes developed as series of parallel lines, each line becoming more specialized but not necessarily more advanced as measured by the criterion of similarity to the next higher class. The growth of the fossil record thus destroyed the idea of a linear progression and, along with it, the oversimplified view that the animal kingdom could be arranged into a well-defined hierarchy.

Attention concentrated so heavily on the continuity of progression because many naturalists feared the implications of a transmutation theory, or, as we should now say, of evolution. But what exactly are we to mean by "evolution" in this context? In the early nineteenth century any theory which connected man with the animals aroused instant hostility, even if it presented the process of development as part of a creative design. But Darwinism has wrought such a change in our perspectives that historians have sometimes been tempted to minimize the significance of the early transmutation theories because they did not anticipate the full range of modern evolutionary philosophy. If we accept Foucault's analysis, referred to above, there could have been no true evolutionism in the eighteenth century, since the most that could be done with classical natural history was to introduce a time scale into the plenum of forms, defining eras in which the successive parts of the whole became manifest. But this is also true of some nineteenth century theories—Chambers' is a good example—and these can hardly be dismissed as of little relevance to the rise of modern evolutionism. Even Darwin admitted that Chambers' book helped to prepare public opinion for the reception of his own theory. Of course Darwinism generated an enormous intellectual revolution beyond the mere introduction of the idea of natural continuity. It broke down the philosophy of design which had pervaded most earlier thought about nature and destroyed the traditional belief that species were natural entities. But as Darwin himself well knew, the fact that the public mind had been "softened up" by earlier, less radical theories of development smoothed the way for his own revolution. The historian of evolutionism cannot dismiss the earlier theories from his consideration simply because they do not fit our modern preconceptions.

Yet in another sense, Foucault's analysis does allow us to make a valuable point. He argues that it was only after Cuvier's reformation of taxonomy that it became possible to distinguish a true history of life; that is, to recognize eras in which life adapted itself to the successive conditions prevailing on the earth. For Foucault the question of transmutation versus the fixity of species is insignificant when compared to the magnitude of this revolution—Cuvier is a founder of modern evolutionism even though he argued against actual transmutation. This approach also allows us to see the work of Cuvier's followers as a real contribution to the rise of the modern viewpoint. Because most of the early nineteenth century paleontologists opposed evolution in favor of successive miraculous creations, it has been all too easy for historians to forget their genuine achievements. John C. Greene, for example, has recently suggested that the theory of successive creations was little more than an extension of the old static world-view, replacing one miracle with many.[12] Yet it was within this creationist system that the early nineteenth century paleontologists mapped out the basic outline of the modern picture of living development. There is some virtue in an approach such as Foucault's, which allows us to see these workers as making a genuine contribution to "evolutionary" thought despite the fact that most of them could not accept actual transmutation. It was not just the early evolutionary speculations that prepared the way for Darwinism, but also the whole system of organic development worked out by the paleontologists who used creationism as a framework within which they could relate the vast array of fossils that was coming to light. Indeed, the destruction of linear progressionism mentioned above is an illustration of the fact that the creationist paradigm actually prepared the way for its own replacement, by allowing the naturalists to uncover trends in the history of life that were more easily explained by Darwinism.

We can also now recognize a distinction within the pre-Darwinian views of living development which cuts across the lines of the transmutation/fixity of species debate that the nineteenth century found so absorbing. As we have seen, many workers at first continued to accept the idea of a hierarchy of organization which in turn defined the progression observed in the fossil record. But they were divided on the cause of the observed progression in a manner which reflects the differing extent to which they had absorbed the implications of the revolution in taxonomy. The explanation of progression put forward by the creationists who followed most closely in Cuvier's footsteps assumed that the sequence of living forms was determined by the directionally changing conditions of the earth's surface. Because of the higher temperature, water level, or carbon dioxide content of the atmosphere (there were a number of alternatives), the early geological periods were thought to have been unsuitable for the higher forms of life. As conditions moved toward those of the present, God was able to create the higher animals, and the history of life thus became a progression. For these workers, the hierarchy of organization did not define the history of creation, it merely served as an indication of the changing needs of adaptation. Geoffroy Saint Hilaire proposed a complementary theory in which the more advanced forms were produced naturally out of the earlier ones by the direct action of the new conditions. Beginning with Louis Agassiz, in particular, however, an alternative view emerged in which the progression was not related to the earth's physical history. Although he knew that the development of life could not be totally independent of the changing climate and so on, Agassiz held that the advancing level of organization was the essence of a distinct plan which God had decided to follow in the order of creation. Man was the goal of this plan, and the progressive sequence thus represented an advance toward the human form that would have occurred whatever the changes in the physical conditions. To some extent, Chambers' theory was a logical extension of this view, in which the plan of development was thought to be programmed into nature so that it unfolded without God's continual interference in the form of miracles. In Foucault's terms, both of these systems represent a retrogression toward an earlier way of thought in which progression can only be the unfolding of a predetermined sequence of forms. The alternative view of successive adaptations, though bound more strongly to the miraculous creation of species, exploited far more completely Cuvier's revolution that allowed organic forms to be seen as adaptive variants of a type rather than as links in a universal chain.

The two views on the ultimate cause of progression differed essentially over the concept of design as applied to nature. Thomas McPherson's recent analysis of the argument from design points out that it can be formulated in two different ways, the first based on design related to purpose, the second on design in the sense of order.[13] The paleontologists who explained the progress of life as a result of changing conditions followed the first of these. Their scientific impetus may have derived from Cuvier, but British naturalists in particular integrated this into the tradition of natural theology propounded by William Paley and the *Bridgewater Treatises*. Here design was seen in its teleological sense, as an effect with a purpose—the adaptation of organisms to their environment was taken as evidence of God's benevolent interest in the well-being of His creation. Progress occurred because He maintained the state of adaptation through new creations as conditions developed toward those of the modern world. For Agassiz and Chambers, on the other hand, design was interpreted in the sense of an order or pattern that could be seen in nature: a transcendental plan in which all forms were related in a comprehensive hierarchy leading up to man. This approach automatically linked man's physical structure into the great design of nature, although Agassiz's commitment to the traditional theology led him to argue that the series of species each must be produced by a miraculous creation, thereby allowing man's distinct spiritual nature to be preserved. Chambers threw off the shackles of the old way of thought to reveal the full logic of the belief that the development of nature (and of mind) follows a predetermined, progressive plan. But many naturalists who were attracted to this concept of design nevertheless shared Agassiz's unwillingness to break with creationism, and some of them went even further than he did in proclaiming the discontinuity of the ascent toward man.

This complex system of technical and theological problems formed the background to the progressionist and transmutationist debates of the early and mid-nineteenth century. But Darwin's theory of evolution by natural selection emerged out of neither tradition, and in fact challenged both concepts of design. In Foucault's terms, Darwin completed the revolution that Cuvier had begun: he not only took adaptation to changing conditions as the basic measure of the history of life, he also elevated it to the status of a driving force that demolished Paley's natural theology by eliminating the need for divine control. Since Darwin also followed Lyell in rejecting the directionalist theories of the earth, his theory could not be expected to generate the kind of progression envisaged by the earlier paleontologists who had related the advance of life shown by the fossil record to changing physical conditions. At the same time, Darwinism made nonsense out of the claim that creation has a particular goal and that the advance of life represents the unfolding of a divine plan. To the extent that he accepted a progressive trend in evolution, Darwin saw it as an indirect and highly irregular by-product of natural selection, an equivalent only of the statistically inevitable progress through diversity postulated in Herbert Spencer's cosmic evolutionism. In this way Darwinism joined with the movement among the paleontologists away from the linear concept of progression. Workers such as Owen had already realized that there was no absolute hierarchy of organization and that the history of life represents a series of branching lines, each moving toward specialization in its own way of life, progression being only an indirect background effect. Darwin capitalized on this movement and provided it with its ultimate justification in the form of his system of completely undesigned development.

The introduction of the Darwinian cosmology represents an enormous change in western man's thought about nature and cannot be understood as a purely technical scientific advance. Yet it is worth remembering that by 1860 the assumptions that underlay the earlier nineteenth century debates were already being undermined by the taxonomists and paleontologists themselves. These workers were finding it increasingly difficult to reconcile the fossil record either with the belief in a series of perfectly designed creations or with the concept of a universal plan aimed at the production of man. At least part of the reason why Darwin's indirect kind of progressionism was able to replace the old forms is to be found in the naturalists' growing sense of the impossibility of reconciling the ever-expanding fossil record with many of their earlier preconceptions.

Notes

[1] On the growing significance of fossils see, for instance, John C. Greene, *The Death of Adam. Evolution and its impact on Western thought*; Francis C. Haber, *The Age of the World, Moses to Darwin*; M.J.S. Rudwick, *The Meaning of Fossils. Episodes in the history of palaeontology*. The original classics in the field are Frank D. Adams, *The Birth and Development of the Geological Sciences*; Sir Archibald Geikie, *The Founders of Geology*; and Karl von Zittel, *History of Geology and Palaeontology*.

[2] For a survey of the literature on this topic see my "Evolutionism in the Enlightenment."

[3] On the growth of the belief in social progress, see J.B. Bury, *The Idea of Progress*, and the work of the same title by Sydney Pollard.

[4] See the edition of Buffon's *Les Epoques de la nature* (1778) edited and introduced by Jacques Roger.

[5] Arthur O. Lovejoy, *The Great Chain of Being. A study in the history of an idea*, lecture IX.

[6] Charles Bonnet, *Palingénésie philosophique, ou idées sur l'état passé et sur l'état futur des êtres vivans, Oeuvres d'histoire naturelle et de philosophie*, XVI, p. 74. Translations here and following are my own, except where another source is cited.

[7] See J.P.B.A. de Monet, chevalier de Lamarck, *Philosophie zoologique, ou exposition des considérations relatives à l'histoire naturelle des animaux . . .*, I, p. 5 and II, p. 49.

[8] On Lamarck's relationship to the idea of an organic series see Henri Daudin, *Etudes d'histoire des sciences naturelles*, II, *Cuvier et Lamarck, les classes zoologiques et l'idée de série animale*, part 2, pp. 110–125. J. Schiller has recently argued that Lamarck did not get his idea of a series from Bonnet's chain of being; see his "L'échelle des êtres chez Lamarck."

[9] See Lamarck, *Hydrogeology*, translated by Albert V. Carozzi.

[10] See R. Hooykaas, *Natural Law and Divine Miracle. The principle of uniformity in geology, biology and theology* (also issued in a revised form under the title *Continuité et discontinuité en géologie*, 1970). See also the same author's articles: "The parallel between the history of the earth and the history of the animal world," and "Geological uniformitarianism and evolution." In addition to Rudwick's *The Meaning of Fossils* see his articles: "The strategy of Lyell's *Principles of geology*," and "Uniformity and progression: reflections on the structure of geological theory in the age of Lyell."

[11] Michel Foucault, *The Order of Things. An archaeology of the human sciences*.

[12] John C. Greene, "The Kuhnian paradigm and the Darwinian revolution in natural history."

[13] Thomas McPherson, *The Argument from Design*; see for instance pp. 6–9.

2

Early Nineteenth Century Progressionism

Enlightenment progressionism had emerged out of the often strange philosophical and theological systems of writers such as Robinet and Bonnet. It had not been based on significant fossil evidence, and it was only in the early decades of the nineteenth century that potentially suitable evidence began to come to light. But the intellectual climate within which these early paleontological researches were conducted was far different from that of the Enlightenment. The late eighteenth century had seen a widespread rejection of James Hutton's uniformitarianism based, in Britain at least, on emotions arising out of a new concern for the Biblical story of creation.[1] The "Neptunist" theory of Abraham Gottlob Werner was taken up as an ideal vehicle for demonstrating that the earth was *not* immeasurably ancient and that its surface has been continually shaped by forces designed to fit it for the eventual creation of man. The rise of what was soon to be called "catastrophist" geology marked a decline in the popularity of Werner's retreating-ocean theory, but retained two of his followers' fundamental beliefs. The earth's surface, they held, has been shaped by violent catastrophes—the last of these being the Noachian deluge—and the occurrence of these catastrophes was linked with a purely directional view of the planet's development. The new directionalism, however, came increasingly to depend on the cooling-earth theory, which was supported by J.B.J. Fourier's study of heat flow and made the basis of the cataclysmic theory of mountain building proposed by Elie de Beaumont (1829–30).[2]

Since the founder of the new science of vertebrate paleontology was himself a leading catastrophist, it is hardly surprising that the two studies were connected from the beginning. When Georges Cuvier began his investigations of the remains contained in the rocks of the Paris basin he was able to point not only to the apparently sudden breaks between the strata but also to certain significant changes in the nature of the successive populations. While the Tertiary rocks contained numerous mammalian fossils, the Secondary series beneath showed no sign of a vertebrate form higher than a reptile. Here was a

15

real basis on which a theory of progression could be erected, and more evidence of progressive steps was produced as other geologists began to extend Cuvier's methods to the earlier stages of the earth's history. Uniformitarianism appeared to be confounded by the clear evidence of a direction existing in the history of life, presumably linked to the earth's physical development.

Two points must be emphasized about this early progressionism, however. The first is the extremely limited and discontinuous nature of the evidence, especially in the first three decades of the century. For this reason alone it is necessary to test with some caution the opinion expressed by historians such as Greene, Eiseley, and Cannon that the catastrophists rapidly became committed to progressionism as an integral part of their world view.[3] In fact, it is difficult to find more than a handful of geologists who made open declarations of support for progression during these early decades. But this cautious attitude reflects more than the imperfect state of the evidence. The catastrophists' real concern was their directionalist theory of the earth, which in turn implied a directionalist, but not necessarily a progressive interpretation of the history of life. An advance in the level of living organization was merely one possible indication of the broader directional changes. Thus even when the progression was recognized, it was treated as a by-product of the earth's physical development. Loren Eiseley has spoken of Charles Lyell's growing opposition to the "transcendental, man-centered progressionism" of the catastrophists, but this gives a seriously misleading view both of the popularity and the nature of the early theory of progression.[4] Unlike Louis Agassiz and a number of later naturalists, the original catastrophists did not explain progression in terms of the unfolding of a transcendental, man-centered plan which the Deity had decided to follow independently of the changing climate. They saw it largely as a result of the directionally changing conditions, which necessarily imposed limits on the order in which the Creator could introduce the different forms of life.

Georges Cuvier and the First Evidence for Progression

Georges Cuvier's pioneering work in vertebrate paleontology certainly provided the first real evidence for progressive steps in the history of life. Yet as a measure of the early caution on this topic, it is significant that Cuvier himself made no effort to suggest that the sequence of populations he discovered was governed by an ascent toward higher levels of organization. Born in 1769, Cuvier was soon drawn to the study of natural history and was brought to Paris by Geoffroy Saint Hilaire. There he gained a degree of scientific and political prestige which allowed him eventually to overshadow both Lamarck and Geoffroy himself. Having perfected the techniques of comparative anatomy, he soon began to apply them to the reconstruction of the fossil bones that were just then being recognized as significant clues to the earth's past. Out of the correlation of the reconstructed forms with the sequence of deposits in which they were found emerged the first real outline of the history of life. In 1811 Cuvier collaborated with Alexandre Brongniart in the production of an *Essai sur la géographie minéralogique des environs de Paris* which established the sequence of strata within the Tertiary, largely on the basis of the invertebrate fossils studied by Brongniart. In the following year Cuvier collected together his own papers on vertebrate paleontology to form his *Recherches sur les ossemens fossiles*. In the *Discours préliminaire* to this work—later published separately under the title *Discours sur les révolutions du surface de la globe*—he combined the stratigraphical technique with his paleontology to present an outline of the succession of vertebrate populations. The *Discours* was rapidly translated into English and continued to be printed, largely unchanged in substance, until after Cuvier's death in 1832. It thus presented a consistent teaching to both the French and British publics over a long period of time, and its influence (or, rather, lack of it) on the rise of progressionism cannot be ignored.

Excellent descriptions of Cuvier's work have been given by authors such as Greene and Coleman;[5] his discoveries can be outlined briefly as follows. First, he argued that there were no known fossil remains of man—evidence that was taken by most other workers as an indication that the human race was a recent creation. Stretching down through the Tertiary he found a series of different populations of mammals, beginning in the more recent deposits with forms closely related to those of today and ending in the most ancient with forms such as *Palaeotherium* which are far removed from any modern genus. (See Plate I.) Beneath the Tertiary was a series of strata containing the remains of reptiles but not of mammals. "It is," he wrote ". . . clearly ascertained that the remains of the oviparous quadrupeds [reptiles] are found considerably earlier, or in more ancient strata, than those of the viviparous class [mammals]."[6] There was, in fact, an enormous discontinuity here between the Secondary series, bare (or as we shall see almost bare) of mammals, and the Tertiary which swarmed with a multitude of different mammalian forms. There appeared to be an equally sudden discontinuity at the beginning of the Secondary, when both the reptiles and the fishes appeared. Since the Transition rocks were at first thought to contain only invertebrate remains, Cuvier was "led to conclude that the oviparous quadrupeds began to exist along with the fishes, and at the commencement of the period which produced the secondary formations."[7] In later editions of the *Discours* he admitted that there were "perhaps even the bones and skeletons of fish" in the Transition rocks,[8] but he retained the remarks implying that the reptiles and fishes appeared together in the Secondary and seems to have had little interest in establishing the existence of an "Age of Fishes" preceding the "Age of Reptiles."

Since Cuvier paid little attention to the early part of the history of life, his discussion contained three points that could be given a progressionist interpretation. These were (1) the late appearance of man, (2) the gradual approach of the mammals toward the modern forms in the Tertiary, and (3) the sudden appearance of the mammals at the beginning of that era. It can easily be demonstrated, however, that he had no intention of incorporating the first two points into a progressionist theory—in fact he actively denied the most obvious implication of his discoveries by arguing that the sequence of fossils did not correspond to the successive appearances of the forms concerned. Cuvier was profoundly opposed to any kind of transmutation theory, but unlike most of his contemporaries he also distrusted the theory of successive miraculous creations. Although he argued that the forms were wiped out by catastrophes at the end of the periods in which their fossils were deposited, he claimed that the appearance of new species after each catastrophe indicated only that they had migrated in from other parts of the earth.

> . . . when I endeavour to prove that the rocky strata contain the remains of several genera, and the loose strata those of several species, all of which are not now existing on the face of our globe, I do not pretend that a new creation was required for calling our present races of animals into existence. I only argue that they did not anciently occupy the same places, and that they must have come from some other part of the globe.[9]

He even applied the same argument to man himself.[10] On this basis it would have been impossible to construct a theory of progression in the Tertiary, since the whole development revealed by the fossil record was in a sense illusory. The only point at which (by implication at least) Cuvier left room for a creation of new forms was at the beginning of the Tertiary when the mammals first appeared.

Although many later writers took the appearance of the mammals as a progressive step in the history of life, there is no evidence that Cuvier himself wished to emphasize this aspect of the process. In general he was profoundly suspicious of the hierarchical view of the organic world which treated one class as obviously superior to another. His division of the animal kingdom into four types or *embranchements*—the vertebrates, molluscs, articulates, and radiates—had destroyed all hope of creating a linear "chain-of-being," and his great *Règne animal* expressed open suspicion of the possibility that a meaningful hierarchy could be constructed among the classes that made up the vertebrate type.[11] With such a background in taxonomy it is hardly surprising that he made no effort to stress that the appearance of the mammals counted as a progression. He would certainly have distrusted any attempt to treat the history of life as though it were governed by a specifically progressive trend. In fact Cuvier offered no consistent explanation of the succession of forms, although there is indirect evidence that he connected it with the changing physical conditions of the earth, in particular with his own early interest in the Wernerian theory of the retreating ocean.[12] Such a view would not have been inconsistent with his belief that it is necessary to understand the "conditions of existence" governing the structure of each organic form, including its relationship to the environment.

The connection with the directional theory of the earth emerges because Cuvier hints that the Tertiary mammals were the first animals adapted to a purely terrestrial habitat. We know that he believed in a steady decline in the sea level, and at one point he argued that when life first appeared there would only have been a few islands standing above the surface of the ocean.[13] The belief that large continents only appeared later in the earth's history may thus explain the unusual use of the phrase "land quadrupeds" in the following passage, a continuation of a quotation given above.

> . . . we are led to conclude that the oviparous quadrupeds began to exist along with the fishes, and at the commencement of the period which produced the secondary formations, while the land quadrupeds did not appear until long afterwards.[14]

In the context, "land quadrupeds" almost certainly means the Tertiary mammals, and the fact that Cuvier describes them in this way (not as "viviparous quadrupeds") suggests that he may have thought of them as the first animals adapted to the large areas of land which appeared as the seas retreated. Such a view would not at first have conflicted with the evidence. Later paleontologists developed the idea of an "Age of Reptiles" in which this class had been dominant both on land and in the sea, but when Cuvier wrote the *Discours* almost all of the known fossil reptiles were aquatic. Take, for instance, his full statement of the principle that the reptiles are more ancient than the mammals:

> It is, in the first place, clearly ascertained that remains of the oviparous quadrupeds are found considerably earlier, or in more ancient strata, than those of the viviparous class. Thus the crocodiles of Honfleur and of England are found underneath the chalk. The *monitors* of Thuringia would be still more ancient, if, according to the Wernerian school, the copper-slate in which they are contained, along with a great number of fishes supposed to have belonged to fresh water, is to be placed among the most ancient strata of the secondary formations. The great alligators, or crocodiles, and the tortoises of Maestricht, are found in the chalk formations; but these are both marine animals.[15]

All of the reptiles mentioned here are aquatic or have a life style dependent on a close association with water. The crocodiles of Honfleur and England shared the characteristics of their modern counterparts.[16] The monitor of Thuringia was a salamander-like creature of aquatic habits[17] and Cuvier specifically noted that the creatures of the Cretaceous Maestricht beds (of which the carnivorous *Mosasaurus* was the most well known) were marine. The only exception known at the time was *Pterodactylus*, a reptile unlike all others in its adaptation for flying.[18] The discovery of *Ichthyosaurus* and *Plesiosaurus* merely added to the number of aquatic reptiles, and it was only in 1824 that the first land dinosaur, *Megalosaurus*, was described. These reptiles were incorporated into the body of the *Ossemens fossiles*, but Cuvier did not acknowledge them in the *Discours*. His general survey of the history of life thus remained consistent with the belief that the reptiles of the Secondary were adapted to a marine environment, or at the most to low-lying or swampy ground. The Tertiary mammals still appeared as the first animals properly adapted to dry land.

Even before the identification of *Megalosaurus*, another discovery had challenged Cuvier's original principles in a far more conspicuous manner. In 1818 a jaw bone was brought to his attention which had been discovered four years earlier by William J. Broderip in the Oolitic slate of Stonesfield, Oxfordshire. He declared it to belong to a species of *Didelphis*, an opossum-like marsupial,[19] and a brief description of the bone was provided in William Buckland's 1824 paper on

Megalosaurus.[20] (See Plate II.) As an obvious exception to the rule that mammalian remains are not found beneath the Tertiary, the case was not allowed to go unchallenged by Cuvier's supporters. Constant Prévost visited England in 1824 and subsequently published a paper questioning the location of the Stonesfield deposits within the Oolitic. Hooykaas has suggested that he was motivated by a desire to defend the theory of progressive development, but the paper does not really support this interpretation.[21] Prévost certainly *was* concerned to re-establish the original rule if possible, but like Cuvier himself he made no effort to stress the element of progression. The discovery could in any case have been incorporated quite easily into a simple hierarchy of ascent, since it was widely admitted that the marsupials were the lowest mammals. Prévost merely pointed out that even W.D. Conybeare and William Phillips' standard *Outline of the Geology of England and Wales* admitted some doubt as to the nature of the Stonesfield beds.[22] Further work soon confirmed, however, that the rocks were Oolitic, and it was some time before another attempt to challenge the counter-example was made. In the meantime Broderip rediscovered another bone unearthed along with the first and then mislaid, describing it as a different but related mammalian form. In 1838 H.D. de Blainville tried to argue that neither of the bones in fact belonged to mammals.[23] He was opposed in France by Achille Valenciennes,[24] and the debate was finally settled in favor of the mammalian interpretation by Richard Owen, who rechristened the two species *Amphitherium broderipi* and *Phascolotherium bucklandi.*[25] The whole debate generated considerable doubt as to when the mammals might first have been created, undermining what had originally been one of the most clear-cut steps in the history of life.

Cuvier made no attempt to discuss the implications of the Secondary mammals in the *Discours*. They would not in any case have undermined a theory which related the history of life to changing physical conditions, since a few mammals might naturally be expected to appear as soon as the environment began to approximate that of the Tertiary. Under the circumstances, however, we can readily understand Cuvier's reluctance to make any open reference to the causes responsible for the late appearance of the mammals, or for any other step in the development of life. Not only was the fossil record highly discontinuous, but no generalization seemed safe against the anomalies produced by the ever-expanding number of discoveries. The Oolitic mammals were at first only a disturbing factor which blurred the clear-cut outlines of Cuvier's original sequence without helping to produce

an alternative interpretation such as a law of gradual progression. Even
if these early "marsupials" were regarded as intermediate in status
between the reptiles and the placental mammals, they could hardly be
seen as a historical link since they were separated from the main
development of the higher class by the whole width of the Cretaceous
system. Cuvier himself would in any case have had little interest in
formulating a progressive system. But even those who retained the idea
of a hierarchy of organization could see that the fossil evidence favored
at best only a highly discontinuous ascent. There was little to suggest
the possibility of a specifically progressive trend in nature, and
attention inevitably remained focussed on external conditions as the
only factor that could be responsible for the major changes between
successive populations.

Adolphe Brongniart
and Geoffroy Saint Hilaire

French natural history in the 1830s was split in two by the debate
that arose out of Cuvier's attempt to crush the transcendental
approach to comparative anatomy being promoted by Geoffroy Saint
Hilaire and his followers. Cuvier's more pragmatic approach took for
granted the adaptation of species to the external conditions and left
their origin a mystery. Geoffroy treated all species as manifestations of
an archetypical form and was willing to assume that new conditions
might evoke a new mode of embryological growth and transform one
species into another. Yet both sides of this great debate accepted
certain limiting ideas which imposed a degree of superficial unity on
their attitudes toward the advance of life. Neither side treated the late
appearance of the mammals as anything but a by-product of the
changing conditions, although there was total disagreement over the
way in which these changes acted. The alternative viewpoints may be
seen most clearly in the work of Adolphe Brongniart and Geoffroy
himself. Brongniart followed Cuvier in refusing to speculate about the
origin of species, but made a notable addition to his theory of the earth
by producing a concrete explanation of how changing conditions could
have allowed the creation of higher forms of life only in the later
geological periods. Geoffroy followed the logic of his own system, but
invoked the same changes in the environment to explain the later
production of the higher forms.

Adolphe Brongniart was the son of Cuvier's co-worker on the
Paris basin. He followed in his father's footsteps, but turned to the

study of fossil vegetables and became the virtual founder of paleobotany through the publication of the first volume of his *Histoire des végétaux fossiles* in 1828. Since plants are more dependent on their environment than animals, Brongniart felt that his work should be of great value in determining the course of the earth's general development.[26] By surveying the plant fossils of the various periods he tried to reconstruct the physical history of the earth, from which he was able to suggest a possible explanation of the sequence of animal populations. Although he shared Cuvier's reluctance to talk about miraculous creation, he seems to have accepted that new forms were produced in this manner, designed to fit the conditions obtaining at the time. The theory developed in his *Prodrome d'une histoire des végétaux fossiles* of 1828 implied that the introduction of the higher forms of life was only allowed when the directional development of the earth brought conditions closer to those of the present time.

Brongniart began by commenting on the fact that many of the fossil plants discovered in Europe resembled those that are still found growing in the equatorial regions.[27] His method of trying to understand the earth's history was essentially an extension of this point, comparing the plants of each period with those of the different geographical regions of today. He distinguished three separate periods before the Tertiary. The first (corresponding to the Carboniferous) contained primitive cryptogamous plants such as ferns. (See Plate III.) Since similar forms are now found growing on tropical islands, Brongniart deduced that both the ancient sea level and the temperature must have been higher than they now are.[28] He also noted that the structure of these early plants suggested that the carbon dioxide content of the atmosphere may also have been much higher in the earlier periods.[29] The next two periods, corresponding to the New Red Sandstone and the Lias, exhibited decreased proportions of cryptogamous plants and a corresponding increase in the numbers of gymnosperms (pines, firs, etc.). Only in the Tertiary did the higher dicotelydons begin to predominate in the population, a development that Brongniart connected with the rise of the mammals at the same time.[30]

On surveying the changes which have taken place in the history of life, Brongniart noted that in both the animal and the vegetable kingdoms the higher organisms only seem to have appeared toward the end of the process. At one point he even hinted that the sequence of creation might actually have been governed by some kind of specifically progressive trend.

We must thus admit that among the vegetables, as among the animals, the most simple beings have preceded the more complicated, and that nature has successively created beings of greater and greater perfection.[31]

But in the end the system he proposed treated the progressive development as a symptom rather than as a cause of the changes in the two kingdoms. Nor was there a strong element of continuity in the sequence of creation, as the above quotation seems to imply. In the vegetables, Brongniart confined himself to four periods, recognizing no gradations within them. In the animals only two steps were involved— the appearance of the reptiles and of the mammals. The idea of nature "successively creating" higher beings reduced itself, in the animal kingdom at least, to two distinct progressive leaps in the level of organization. Cuvier had established the sudden appearance of the mammals and the possibility that the reptiles had appeared later than the fishes was now beginning to appear. Unlike Cuvier himself, Brongniart accepted the idea that these steps really did constitute a progression, but the theory that he proposed was merely an attempt to account for the steps in terms of the changing physical conditions.

The basic assumption of the *Prodrome* was that the changes in the climate were themselves responsible for the vegetable progression. As the sea level, temperature, and carbon dioxide content of the atmosphere declined, so it became possible for the higher plants to be introduced. In the same year Brongniart published a paper in which he extended this idea to provide an explanation of the animal progression. He now argued that the developments in the vegetable kingdom were actually responsible for producing the reduction in the carbon dioxide level, and that it was this factor which allowed the higher animals to be created in later periods. Speaking of the Carboniferous, he wrote:

> During this first period, the atmosphere was freed of a part of its excess of carbon, by the vegetables which grew upon the land, which assimilated it, and were afterwards buried in the state of coal in the bowels of the earth. It is after this epoch, during our second and third periods, that the immense variety of monstrous reptiles began to appear, animals which, by their mode of respiration, are yet capable of living in a much less pure air than that which the warm-blooded animals require, and which, in fact, have preceded them at the earth's surface.
>
> The vegetables continued to withdraw a part of the carbon from the air, and thus rendered our atmosphere daily more pure; but it was only after the appearance of a quite new vegetation, rich in large trees, and the origin of the numerous deposits of lignite, a vegetation which appears to have covered the surface of the earth with vast forests, that a great

number of mammiferous animals, resembling in the essential features of
their organization, those which still exist upon the earth, for the first time
made their appearance upon its surface.[32]

Here was an ingenious explanation that reduced the progressive steps
in the history of the animal kingdom to by-products of the earth's
physical development. Progression was certainly not treated as a
programmed ascent through a hierarchical plan of creation, and it is
significant that Brongniart made no mention of the last appearance of
man—a central point of later, more explicitly progressionist theories.
Indeed, there was no way in which he could have accounted for this
last step. Relating the development of the animal kingdom to the
decline of the carbon dioxide level could explain why a particular class
such as the mammals could only emerge at a later date, but it could not
explain why one mammalian species should appear any earlier or later
than the others. The theory explained the broadest of the discontinuities revealed by Cuvier's researches, and in thus rationalizing the
step-by-step introduction of the classes it probably reflects an implicit
belief in miraculous creations. But Brongniart's refusal to speculate
openly about the miraculous introduction of new species reflects the
lack of theological interest that was typical of many French naturalists.
The idea of a divinely planned ascent toward man would not have
occurred to such workers; the alternative of relating the ascent to the
earth's physical development would have seemed far more plausible.

If Brongniart's system merely developed principles inherent in
Cuvier's approach, there was an alternative movement underway to
explore the possibility that changing physical conditions might be the
direct cause of the appearance of new forms. Etienne Geoffroy Saint
Hilaire proposed a transmutation theory embodying this point as part
of his campaign to challenge Cuvier's whole program of natural
history. Since Geoffroy held that transmutation was caused by the
external conditions his theory has sometimes been compared with that
of Lamarck, but in fact there is little resemblance. Where Lamarck
introduced an additional force constantly tending to push organisms
further up the scale of complexity, Geoffroy ignored the hierarchical
view of nature and regarded transmutation as a sudden effect by which
another potential variant on the basic plan of all living structures was
produced.[33] His study of comparative anatomy embodied a search for
the archetypical form upon which all the vertebrates were modeled.
This view was proposed in his *Philosophie anatomique* of 1818–22 and
in later years Geoffroy and his followers tried to extend the system so
that other types could be admitted as variants of the same plan. The

theory of transmutation formed a basic part of the system, explaining how the various manifestations of the archetype were produced in response to changing conditions. Cuvier's efforts were devoted to a far more pragmatic study of the relationships between living forms, and he inevitably rejected both the transmutation theory and the whole system of transcendental anatomy on which it was based.[34]

Despite the difference between their fundamental philosophies, Geoffroy retained some of the superficial characters of the system outlined by Cuvier and Brongniart. He accepted that the late appearance of the warm-blooded animals constituted a progression[35] and continued to treat it as a discontinuous step in the history of life, the result of a sudden mutation of the original living forms. He also agreed that the cause of the development must be sought in the changing physical conditions—Geoffroy had no more interest than Brongniart in a specifically progressive plan of creation. He did at one point compare the development of life to the transformation of the fish-like tadpole into the frog, and this has led Hooykaas to argue that he *did* see an analogy between the history of life and the predetermined development of the embryo.[36] Such a comparison between ontogeny and phylogeny—the so-called recapitulation theory—certainly appealed to later progressionists. One of Geoffroy's followers, E.R.A. Serres, suggested in his *Anatomie comparée du cerveau* (1824–26) that the mammalian embryo passes successively through phases corresponding to the sequence of the lower classes.[37] But Serres himself did not relate this sequence to the fossil record, and Geoffroy's own embryology makes it extremely unlikely that he could have appreciated such an analogy. In the *Philosophie anatomique* he argued that each individual vertebrate is provided with structures which can develop it either for air or water breathing.[38] The case of the tadpole was seen as the disappearance of one of these sets of organs and their replacement by the other, a process which Geoffroy took to be similar to that by which monstrosities are produced.[39] He almost certainly saw the first appearance of the warm-blooded animals as a similar kind of modification, with the emphasis not on the programmed development of higher forms but on the elucidation of alternative structures by changing conditions.

Geoffroy's transmutation theory was first hinted at during his study of an extinct crocodile-like form which he named *Teleosaurus*. He was convinced that these animals had lived under conditions quite different from those of today and suggested that the changing environment was directly responsible not only for the mutation of

Teleosaurus into the modern crocodile but also for the appearance of the warm-blooded animals.[40] This idea was developed further in a paper entitled "Le degré d'influence du monde ambiant pour modifier les formes animales" published in 1833. Here he introduced his embryological studies, arguing that each individual organism would always develop like its parent unless the external conditions changed— clear evidence that he postulated no inherently progressive trend.[41] Noting that the embryo of a bird has some resemblance to that of a reptile he argued that to change the one into the other would require only some accident "quite inconsiderable in its original production" during the early stages of growth.[42] The accident would of course be provided by the changes in the earth's physical state; Geoffroy made no detailed study of geology and merely confined himself to speculating that the crucial change would be a modification in the state of the atmosphere.[43] As the conditions began to change toward those of today, a few individuals would develop with the higher structure, apparently as monstrosities. Eventually, however, however, more and more individuals would develop in the new way and at last the whole species would be converted. The warm-blooded animals were thus produced from the reptiles as a sort of "collective monstrosity" when the new conditions had triggered off the growth of the new respiratory structure in the whole population.

Geoffroy's theory was highly speculative and was denounced as such by Cuvier and his followers, who preferred to believe in fixed species of unknown origin. The influence of the theory was thus limited, as indeed was the influence of its parent system, Geoffroy's transcendental anatomy. Yet the attempt to explain the origin of new classes by transmutation shows that both sides of the great debate accepted certain basic points, which thus formed a common frame-work within which all French naturalists viewed the history of life on the earth. Cuvier's discovery of the sudden appearance of the mammals in the Tertiary came to be regarded as evidence for a progressive development of life, but the complete discontinuity of the step was universally accepted—not even Geoffroy wanted to derive the higher animals gradually from the lower ones. The cause of the development was sought only in the changing conditions, seen either as a direct influence on living forms or as a limit on the sequence of creation. Neither Geoffroy's transcendentalism nor the more pragmatic ap-proach of Cuvier and Brongniart allowed the resurgence of any equivalent to Lamarck's inherent progressive trend in the history of life, nor did the fossil evidence at this time give any support to such a belief. The universe *was* a directional system, and it was widely accepted that the nature of the earth's development had ensured that as a matter of fact the higher animals could only appear at a later date. But there was no necessarily progressive plan of creation, least of all aimed at the eventual production of man.

Catastrophism in Britain

The Franch naturalists may have been indifferent to theological issues, but their British counterparts were passionately concerned that the growth of the new science should not interfere too obviously with the orthodox view of nature. This concern—described in detail by writers such as Gillispie and Haber—imposed a number of restrictions on their thinking. It considerably retarded acceptance of the possibility that the earth might be very ancient, and even when this point began to be admitted the choice of suitable theories was severely limited. Werner's theory was the vehicle through which the traditional view of creation was at least partially retained, since it demanded only a limited time scale. As Neptunism was gradually replaced by the more comprehensive system of catastrophism the time scale was gradually expanded, but it continued to be accepted that the earth's history was directional—it had a beginning and an end, as implied by Genesis. To some extent understanding of the history of life was modified by the need for reconciliation with the sequence of creation derived from the Bible. In the case of at least one important writer—James Parkinson— this requirement led to a rather strange approach to progressionism. Finally, concern for the traditional view promoted the belief in past catastrophes, and in particular the identification of the last upheaval with the Noachian deluge. This gave rise to the whole school of "Diluvialist" geology best represented by William Buckland's *Reliquiae Diluvianae* of 1823.

The influence of the theological concerns should not, however, be overemphasized, nor should it be regarded as a totally negative factor. Once the more literal interpretations of Genesis had been dispensed with, there were a number of fronts along which the catastrophists were able to advance their science. As Walter Cannon has pointed out, even geologists such as Buckland were not as orthodox as some historians have imagined.[44] Their efforts to work out a system that was both scientifically and theologically acceptable led to notable expansions of the fossil record which had enormous impact on our understanding of the history of life. Since they were only too eager to work with the idea of miraculous creation, the catastrophists never tried to avoid the implications of the sequence of fossil deposits as Cuvier had done. They realized from the beginning that there has been a real sequence of distinct populations, a point that was of considerable value to the more pious workers who followed in the essentially pragmatic footsteps of William Smith to develop and extend the system of fossil-based stratigraphy. Workers such as Adam Sedgwick and Roderick I. Murchison created a series of geological periods extending far below those that had been recognized by Cuvier, while even Buckland made notable additions to our knowledge of the vertebrate populations in the antediluvian periods. Acceptance of miraculous creations promoted a new interest in natural rather than revealed theology as a guide to geological research. As Buckland pointed out in his *Vindiciae Geologicae* (his inaugural address at Oxford, published in 1820), the system of thought proposed in William Paley's *Natural Theology* could be greatly extended by showing that the same degree of divine benevolence governed each of the earlier populations.[45] The paleontologist was thus led to study the structure of the fossils he discovered in connection with the external conditions of the period in which the creatures had lived, to establish the adaptation of structure to function which Paley took as almost the sole evidence of design.

The concern for this kind of natural theology inevitably blunted some implications of the Genesis story in a manner that has particular relevance for the early history of progressionism. The catastrophists were inevitably directionalists and they saw the purpose of the whole creation as the gradual preparation of the earth for the appearance of man. But by 1830 it was firmly accepted in responsible geological circles that divine providence was manifested in the physical world not through continual miracles but in the original design of the system itself. God had created the world in such a way that it would only gradually become ready for man, and the paleontologist could argue that His great benevolence has ensured that at each stage in the process the planet has been inhabited by a population specially designed and created to suit the conditions of the time. No one would have considered this as the imposition of a limit on the Creator's powers— after all, He was free to choose the nature of the physical universe, after which one would expect a rational God to restrain His benevolence within the range of what was consistent with the physical development. Natural theology thus tied in with directionalism to give an explanation of the progressive steps observed in the fossil record which was essentially similar to that proposed by Brongniart. There was no need to assume that there was a specifically progressive plan of organic creation through which life was made to ascend a transcendental hierarchy of organization toward the human form. The whole sequence could be explained as a natural result of the physical developments, with revelation merely confirming that the final appearance of man was the fulfillment of the whole process. The organic progression was only a by-product of the physical directionalism and hence received comparatively little emphasis. Lyell's attack on the concept in 1830 should not mislead us into thinking that the necessity of an advance toward higher forms was an integral part of the catastrophist position—he had his own reasons for being sensitive on that particular issue.

The comparative lack of emphasis on progression was a result both of its secondary importance and of the continuing irregularity of the fossil evidence itself. Although the range of the fossil record was greatly extended and new progressive steps added, the evidence still did not suggest that the general sequence of populations was governed by a progressive trend. There was no continuity of development—on the contrary, the advance of life occurred by means of great irregularities which fitted naturally into a system where miraculous creation was combined with the belief that the greater intensity of past geological agents had resulted in vast catastrophes. As long as there was no direct theoretical reason to suppose that the progression was the most significant factor in the development of life, the evidence itself would hardly suggest such an interpretation, even when it did indicate that there had been a number of progressive leaps in the level of organization.

Attempts to argue that the earth's history was essentially compatible with the sequence of events described in Genesis had been popular among the early Wernerians, the work of J.A. Deluc being perhaps the best example. But as the nineteenth century advanced it gradually became obvious that the fossil record was becoming more difficult to reconcile with the scriptures. The sequence of events described in Genesis I is as follows: on the third day of creation dry land was formed and populated immediately by vegetable life; on the fifth day marine life and birds were created, and on the sixth day the land animals were formed concluding with man himself. Even accepting that each "day" could be regarded as a long period, there were problems here for the geologist. The earliest known fossils were the invertebrates of the Transition rocks, not vegetables. The appearance of the mammals—Cuvier's greatest discovery—was simply ignored by Genesis, although the apparent absence of terrestrial reptiles in the Secondary rocks at first allowed the mammals to be treated as the earliest land animals. Only on the question of the final appearance of man did the fossil record and the Good Book agree completely, and it was this point that most easily allowed the sequence to be regarded as a progressive one.

One of the last serious attempts to defend a scriptural interpretation of the fossil record was made by James Parkinson in his *Organic Remains of a Former World* (1804–11). Parkinson was a surgeon and at first his geological opinions were rather out of date, although his fossil descriptions are still regarded as important. He even suggested that the coal measures were formed when the deluge covered large numbers of antediluvian trees with mud.[46] But in preparation for his last volume on the vertebrates he read Cuvier's papers and absorbed a more sophisticated viewpoint. He now accepted that there has been a whole series of different creations corresponding to the various strata, although the sequence itself could still be reconciled with Genesis. Ignoring the fact that the vegetables ought to come first, he pointed out that the coal deposits contained the remains of both plants and marine invertebrates, after which the sequence was exactly as revealed to us.

> In all these strata [immediately above the coal measures] no remains are to be found but those of the inhabitants of the waters; excepting those of birds, which exist, though rarely, in some particular parts. But in none of these strata has a single relic been met with which can be supposed to have belonged to any terrestrial animal.
>
> In the next period it is stated, the beasts of the earth, cattle and everything that creepeth upon the earth, were made. The agreement of the situations in which the remains of the land animals are found with the stated order of creation, is exceedingly exact; since it is only at the surface, or in some superficial stratum, or in comparatively some lately formed deposits, that any remains of these animals are to be found.
>
> The creation of man, we are informed, was the work of the last period; and in agreement with his having been created after all the other inhabitants of the earth is the fact that not a single decided fossil relic of man has been discovered.[47]

In order to confirm the early appearance of the birds, Parkinson accepted at their face value certain remains of which Cuvier was extremely suspicious.[48] The basic step he described was from aquatic to terrestrial life, a step which we have already shown to be reasonably consistent with the early very limited knowledge of the Age of Reptiles and which seems to have been supported by Cuvier himself.

An unusual feature of Parkinson's account is his refusal to

connect his ideas with any particular theory of the earth, although they would have fitted in well with the then currently popular Neptunism. Instead he argued that the real pattern of development, both revealed and observed, was that of a progression ending with man.

> Does it not appear from this repeated occurrence of new beings, from the late appearance of the remains of land animals, and from the total absence of the fossil remains of man, that the creative power, as far as respects this planet, has been exercised continually, or at distant periods, and with increasing excellence, in its objects, to a comparatively late period: the last and highest work appearing to be *man*, whose remains have not yet been numbered among the subjects of the mineral kingdom.[49]

Here was a fairly clear statement that there has been a progressive plan of organic creation; yet Parkinson's remarks cannot really be connected with the later development of this belief. For him, progression meant merely the following out of the Genesis sequence toward man—a development from marine to terrestrial life rather than an ascent of the normally accepted hierarchy of the vertebrate classes. Indeed his claim that the birds were created along with the first marine life would have totally destroyed what many progressionists would have regarded as a real sequence of increasing complexity. Only in the final appearance of man did this kind of Biblical progressionism fit in with the later version developed by naturalists such as Louis Agassiz.

Parkinson's position was tenable when it was first proposed, and in fact the same idea that the progression of life supports the Biblical story was advanced in William Kirby's *Bridgewater Treatise* as late as 1835.[50] But by this time the connection had been rendered obsolete by the discoveries of the 1820s and 1830s, which made it clear that the sequence in which the classes were introduced did not correspond to Genesis. To a large degree this extension of the fossil record came about as many geologists took up the stratigraphical techniques of William Smith, the "father of English geology," and applied them to an ever wider range of strata.[51] Smith himself worked with invertebrate fossils, so his work had little direct influence on progressionism. He was in any case far too practically minded to have bothered with such a theoretical issue. But at the same time his followers were not slow to take up Cuvier's technique of discovering and reconstructing the vertebrate remains from the different strata, and in this way important steps were added to the history of life. At first the vertebrates discovered in the Secondary rocks did not undermine Parkinson's Biblical reconciliation. In 1814 Sir Everard Home described the remains of a creature which he subsequently established as a marine reptile; the modern name *Ichthyosaurus* was bestowed on the genus four years later by Koenig.[52] In 1821 W.D. Conybeare and H.T. De la Beche described the partial remains of another marine reptile under the name *Plesiosaurus*. An almost complete skeleton of the same form was discovered a little later and described by Conybeare.[53] (See Plate IV.) These were both marine forms, however; they extended the scope of the Age of Reptiles without breaking down the generalization that all of the early members of this class were aquatic. But in 1824 Buckland described to the Geological Society the remains of a gigantic terrestrial reptile from the Oolitic slate of Stonesfield.[54] He named it *Megalosaurus* and regarded it as a lizard with rather an unusual dental structure. We now know it as a dinosaur—the first to be discovered. A second dinosaur, *Iguanodon*, was reported by G.A. Mantell in the following year[55] (the teeth of this form, resembling those of the modern iguana, had been discovered in 1822 but it was only after some confusion that Mantell established its true nature). In 1831 Mantell published a popular paper under the title "The geological age of reptiles," describing the Secondary era as a time dominated by this class on the land as well as in the sea.[56] The idea of a sea-to-land transition at the beginning of the Tertiary had to be rejected now, with the emphasis being transferred to the essentially progressive step from the Age of Reptiles to the Age of Mammals.

The second major development of these decades was the classification of the hitherto highly confused Transition series. To a large extent this was the work of two men: Adam Sedgwick, professor of geology at Cambridge since 1818, and Roderick I. Murchison, who gained his first experience in geology helping Sedgwick to explore the Old Red Sandstone of Scotland in the late 1820s. In 1831 both men set out separately to investigate the ancient rocks of Wales. Murchison chose the southern counties and after some years traced out a whole series of strata stretching down from the base of the Old Red Sandstone. He named the series after an ancient tribe which had inhabited the area and his classic *Silurian System* appeared in 1839. Sedgwick had in the meantime launched straight into the even older and more complex rocks of north Wales, eventually elucidating their structure under the name of the Cambrian system. It was at first difficult to distinguish the upper reaches of Sedgwick's Cambrian from the lower layers of Murchison's Silurian, a fact which led to a dispute between the workers and an unfortunate estrangement. But before this happened they had studied and finally accepted the conclusion of

William Lonsdale that the rocks of Devon should be associated with the Old Red Sandstone of the north to give another system named the Devonian.[57] A complete sequence was thus established, stretching up from the Cambrian through the Silurian and Devonian to the base of the Carboniferous and the Secondary series.

The significance of these developments for progressionism lay in the distribution of invertebrate and fish remains through the three systems. As early as 1828 Sedgwick and Murchison had reported the discovery of fish in the Caithness schist of the Old Red Sandstone, and by the mid 1830s the Devonian system was known to be well populated with members of this class.[58] Murchison also found fish in the upper strata of the Silurian, but the lower Silurian and all of the Cambrian appeared to contain only the remains of invertebrates such as the Trilobites. (See Plate V.) A wide progressive sequence could now be established from an Age of Invertebrates to an Age of Fishes and then on to the already established Ages of Reptiles and Mammals. In 1841 John Phillips introduced the modern names for the three basic eras of the fossil record: "Palaeozoic" for the Age of Fishes and Inverte-brates, "Mesozoic" for the Age of Reptiles and "Cainozoic" for the Tertiary or Age of Mammals.[59] Although he established the eras on the basis of invertebrate fossils, their convenient identification with the development of vertebrate life helped to emphasize the step-by-step development of life. And as yet the steps were still quite discontinuous. The earliest fish were strange creatures, often heavily armored, but they were by no means primitive examples of their class. There was as yet no sign that amphibians preceded the appearance of the true reptiles. Just as with the sudden flourishing of the mammals at the beginning of the Tertiary, each of the lower classes seemed to begin with the immediate introduction of a number of well-established forms.

These discoveries were spread over several decades and we should hardly be surprised that a fully fledged progressionism did not arise before about 1840. There was, in any case, a perfectly acceptable way of explaining the progressive steps already available—they could simply be incorporated as evidence for the directional modifications of the earth itself. This approach was adopted very early in the century by the British Neptunists, serving as an obvious alternative to Parkinson's Biblical account of the fossil record. Robert Jameson, the leading British follower of Werner and perhaps the least theologically inclined, observed in the third volume of his *System of Mineralogy* (1808) that in the succession of creations both animals and plants increase "in number, variety and perfection."[60] But he connected this quite explicitly with the decline in the water level postulated by Werner, pointing out that the higher animals could not have existed when the dry land was only just beginning to emerge.[61] In later years when he had been forced to abandon Werner's theory for the more general catastrophism, Jameson still kept up his interest in this kind of explanation for the development of life. The notes he added to the later editions of Cuvier's *Essay on the Theory of the Earth* refer briefly to the progression of life, as does an extract from Alexander von Humboldt's *Geognostical Essay on the Superposition of Rocks* which was printed in Jameson's *Edinburgh Philosophical Journal*.[62] The same journal, however, carried a translation of Brongniart's important paper, along with a comment from Jameson claiming that he had been teaching a similar system for some years—clear evidence that he still regarded the progression only as a consequence of changing physical conditions.[63] A brief attempt to propose a more directly progressionist interpretation of the fossil record came in 1826 with Robert Grant's "Observations on the nature and importance of geology," also published in Jameson's journal.[64] But here the evidence for a progressive trend in nature was used to support Lamarck's otherwise completely rejected theory of evolution, and Grant's article was ignored. In general, standard works such as Conybeare and Phillips' *Outline of the Geology of England and Wales* simply avoided the whole issue of progression. Thus, when Charles Lyell came to attack progressionism in the first volume of his *Principles of Geology* (1830), he seems to have been hard pressed to find anyone who had spoken out openly enough in its favor to be cited as an opponent. In the end he turned to Sir Humphrey Davy, who was no geologist, but had given some opinions on the subject in the highly philosophical *Consolations in Travel* which he wrote just before his death. Davy referred to the late appearance of the higher forms on a number of occasions in this work, allowing Lyell to put together a composite quotation which certainly gave the impression that he believed in a progressive trend in creation.[65] A passage which Lyell did not use, however, makes it clear that Davy too related the progressive steps to the changing physical conditions—he argued that the higher animals could not have withstood the excessive violence and heat which prevailed in the earth's early history.[66]

Since Lyell's real attack was launched against the combination of catastrophism and directionalism, the supporters of that viewpoint naturally hastened to its defence. They all wished to retain the steps in the history of life as real evidence of a directional trend, but in some cases they did so without emphasizing the higher state of organization of the later forms. This suggests that they were far more interested in connecting the steps with changing physical conditions than with a specifically progressive organic trend in creation. W.D. Conybeare, for instance, attacked Lyell's work both in the *Philosophical Magazine* and in an 1832 report to the British Association.[67] Although he referred to the differences between the earlier and later populations to support the directional approach against the uniformitarian, he did not stress the element of progression involved. Only in 1841 did he urge the progression of life as an argument against the steady-state world view, in a private letter to Lyell which Rudwick has but recently brought to light.[68] William Whewell—perhaps Lyell's most philosophical opponent—argued strongly for catastrophism in works such as his *History* and *Philosophy of the Inductive Sciences*, yet admitted in a review of the *Principles* that Lyell had "shown . . . that the evidence for what has been called the successive development of life, as derived from the earth's strata, fails altogether."[69] Even Murchison's great *Silurian System*, which has been cited by Cannon as an explicit defence of progressionism, concentrates only on the directional theory and in fact hardly mentions the actual advance of living organization.[70] Once a careful distinction is made between directionalism and progressionism, it becomes clear that for many catastrophists it was the former that was the really important issue, the latter being only a minor piece of evidence.

Two workers who did, however, make definite references to progression in defence of catastrophism were Adam Sedgwick and William Buckland. In 1831 Sedgwick delivered the presidential address to the Geological Society, making use of the opportunity to deliver an attack against Lyell's views which includes the following passage:

> I think that in the repeated and almost entire changes of organic types in the successive formations of the earth—in the absence of mammalia in the older, and their very rare appearance (and then in forms entirely unknown to us) in the newer secondary groups—in the diffusion of warm-blooded quadrupeds (frequently of unknown genera) through the older tertiary systems—in their great abundance (and frequently of known genera) in the upper portions of the same series—and, lastly in the recent appearance of man on the surface of the earth (now universally admitted)—in one word, from all these facts combined, we have a series of proofs the most emphatic and convincing,—that the existing order of nature is not the last of an uninterrupted succession of mere physical

events derived from laws now in daily operation: but on the contrary, that the approach to the present system of things has been gradual, and that there has been a progressive development of organic structure subservient to the purposes of life.[71]

The scope of this account of progression is still rather limited—it includes no reference to the possibility that the Old Red Sandstone which Sedgwick was already exploring might constitute an Age of Fishes. Furthermore, it is most unlikely that Sedgwick intended his remarks to be taken as anything more than support for a directional view of the earth's history. As he said, the progressive development was "subservient to the purposes of life," and his later career shows that he was profoundly opposed to the idea that life has advanced toward man through the unfolding of a transcendental organic plan. He made no further reference to progression until 1845 when he attacked Chambers' *Vestiges of Creation* in a review that seemed almost to deny the advance of life altogether. At the same time he wrote to Agassiz grumbling about the pernicious effects of transcendentalism which, he claimed, shut out the concept of design altogether. For Sedgwick, God's handiwork could be seen only in the adaptation of successive populations to the changing conditions, not in the supposed existence of an overall organic plan. As far as he could see the latter led only toward transmutationism, which of course he abhorred. He seems to have been unable to appreciate that Agassiz at least could accept the existence of such a plan and still believe that it was worked out through a series of miracles. It was thus only in a very restricted sense that he could admit the progression of life.

Now I allow (as all geologists must do) a kind of progressive development. For example, the first fish are below the reptiles; and the first reptiles older than man. I say we have successive forms of animal life adapted to successive conditions (so far proving design), and not derived in natural succession in the ordinary way of generation.[72]

The passage clearly suggests that the overall progression was only grudgingly to be admitted as evidence of the earth's physical development; it was not an autonomous factor in the history of life. It is unlikely that Sedgwick meant his brief remarks in 1831 in anything but the same rather limited sense.

The use of progression as evidence for the directional view of the earth's history is far clearer in William Buckland's *Bridgewater Treatise* of 1836. This series of treatises was commissioned in the will of the eighth Earl of Bridgewater, essentially for the purpose of extending Paley's version of natural theology. Buckland was asked to write the volume on geology, giving him an ideal opportunity to relate each step in the development of life to the changing physical conditions, the adaptation at each stage demonstrating the Creator's benevolence. He accepted directionalism in the form of the cooling-earth theory, which he connected with the nebular hypothesis of the formation of the solar system. As an example of a correlation between this theory and the history of life he mentioned the possibility that the armor and thick scales of many early fishes might have served as protection against the higher temperature of the ancient seas.[73] Similar factors governed the creation of the higher forms in the later periods. We can show, Buckland claimed

> . . . that the creatures from which all these fossils are derived were constructed with a view to the varying conditions of the surface of the earth, and to its gradually increasing capabilities of sustaining more complex forms of organic life, advancing with successive stages of perfection.[74]

For example, the state of affairs at the beginning of the Tertiary had of necessity made the reptiles of the Secondary era extinct, but the conditions were not yet suitable for the modern mammals so the early Tertiary had first been populated by mammals of a kind different to those of today.[75] In the end, Buckland was cautious enough to admit that there has not really been an absolute progression in the level of organization, only a change in the proportion of the classes making up the successive populations. "It is indeed true," he wrote, "that the animals and vegetables of the lower classes prevailed *chiefly* at the commencement of organic life, but they did not prevail exclusively." To back up this statement he described anomalies such as the Stonesfield marsupials and concluded: "Thus it appears that the more perfect forms of animals became gradually more abundant, as we advance from the older series of depositions."[76] This was compatible with the belief that the real driving force behind the sequence of creations was the directionally changing conditions, but a complete negation of the idea that the history of life is governed by a real progressive trend. Buckland went even further and argued that some of the early fishes were so highly organized that "a kind of *retrograde* development, from complex to simple forms, may be said to have taken place."[77] Such remarks clearly illustrate the subordinate nature of progressionism in the catastrophist system.

Even for Buckland and Sedgwick, then, progression was reduced to a highly qualified piece of evidence for directionalism. The possibility of a specifically progressive trend toward man in the sequence of creations was not an integral part of the catastrophist position. By concentrating on natural theology, on the adaptation of each population to its environment, the catastrophists were able to maintain a separation between man and the animal kingdom based on far more than a mere belief in miraculous creation. Revelation might assure us that the world was designed eventually to produce a theatre for the human drama, but the sequence of fossil populations was only indirectly connected with this final act. To accept this interpretation of the catastrophists' position is to see their reaction to Lyell's attack in a new light. Gillispie has argued that in raising progressionism as an argument against Lyell, Sedgwick was moving onto very thin ice, yet at the same time he has expressed surprise that the debate over the *Principles of Geology* was characterized by none of the animosity which greeted Chambers' *Vestiges*.[78] But we should only need to be surprised if the catastrophists were committed to Eiseley's "transcendental, man-centered" progressionism—then Lyell's emphasis on continuity would immediately have raised the specter of a gradual advance that would integrate man into the animal kingdom in a way that everyone found unacceptable. But in fact Sedgwick was standing on much thicker ice than Gillispie imagines, since his kind of progression did not in any case treat man as the goal of a comprehensive plan of organic creation. The final appearance of a being combining moral attributes with the highest known organic structure was not in any sense the last step in the advance of life, to be treated merely as a continuation of the earlier trend. The principle of continuity in the physical world posed no real threat; not just because the progressive steps were so evidently discontinuous but also because the system made no effort to see man as the final product of the organic progression. This is why the debate over the *Principles* was so restrained—the kind of progression supported only half-heartedly by the catastrophists could never have raised the implications later brought out by Chambers, and Lyell went further than anyone else in denying the connection between man and the animals. Nevertheless by 1840 the basic sequence through the vertebrate classes in the fossil record had been established. Not everyone shared the ideals of the tightly knit school of British geologists and it was only a matter of time before a new and more dangerous interpretation would be put on the fossil evidence.

Notes

[1] On these developments see the books by Greene and Haber cited in Chapter one, also C. C. Gillispie, *Genesis and Geology*.

[2] See for instance J.B.J. Fourier, "Mémoire sur les températures du globe terrestre et des espaces planétaires"; and Léonce Elie de Beaumont, "Recherches sur quelque-unes des révolutions de la surface du globe."

[3] See Greene, *The Death of Adam*, p. 125; Loren Eiseley, *Darwin's Century. Evolution and the men who discovered it*, p. 108; Walter F. Cannon, "The uniformitarian-catastrophist debate," and "The problem of miracles in the 1830s." See also Gillispie, *Genesis and Geology*, p. 131; and Hooykaas, *Natural Law and Divine Miracle*, pp. 98–99.

[4] Eiseley, *Darwin's Century*, p. 108.

[5] See Greene, *The Death of Adam*, pp. 89–119; and William Coleman, *Georges Cuvier, Zoologist. A study in the history of evolution theory*, pp. 107–140.

[6] Georges Cuvier, "Discours préliminaire," *Recherches sur les ossemens fossiles de quadrupèdes, ou l'on retablit les charactères de plusieurs espèces d'animaux que les révolutions du globe paroissent avoir disparu* (Paris, 1812), I, p. 68. The English is taken from Robert Kerr's translation: Cuvier, *Essay on the Theory of the Earth*, (1813), p. 106.

[7] *Ossemens fossiles* (1812), I, p. 69; *Essay*, pp. 107–108. See also the fourth edition of the *Ossemens fossiles* (Paris, 1834), I, p. 193. The same conclusion is implied *ibid*, p. 357.

[8] *Ossemens fossiles* (1834), I, pp. 359–360. Coleman, *Georges Cuvier*, p. 128, gives an adaptation of a diagram prepared by Alexander von Humboldt for the later editions of the "Discours" which suggests that the fish do occur before the reptiles, but the original is not so clear on this point.

[9] *Ossemens fossiles* (1812), I, p. 81; *Essay*, pp. 125–126.

[10] *Ossemens fossiles* (1812), I, pp. 84–85; *Essay*, p. 131.

[11] See Cuvier, *Le Règne animal distribué d'après son organization, pour servir de base à l'histoire naturelle des animaux et d'introduction à l'anatomie comparée*, I, pp. xx–xxi. See also Cuvier's earlier paper, "Sur un nouveau rapprochement à établir entre les classes qui composent le règne animal," especially 80–81.

[12] On Cuvier's early Wernerianism see Coleman, *Georges Cuvier*, p. 112.

[13] *Ossemens fossiles* (1812), I, pp. 7–8 and 10; *Essay*, pp. 12–13 and 13–14.

[14] *Ossemens fossiles* (1812), I, p. 69; *Essay*, p. 108.

[15] *Ossemens fossiles* (1812), I, p. 68; *Essay*, pp. 166–167.

[16] For Cuvier's full description of these creatures see *Ossemens fossiles* (1812), IV, "Sur les ossemens fossiles du crocodiles," art. IV, pp. 16–19; art. VI, pp. 31–33 and art. VII, pp. 33–37.

[17] *Ibid.*, art. I, pp. 4–11, see especially p. 6.

[18] For the full description of *Pterodactylus* see *ibid.*, IV, "Sur quelques ovipaires fossiles des schistes calcaires," art. V, pp. 24–37.

[19] Cuvier recorded his opinion in a footnote added to later editions of the *Ossemens fossiles*; see edition of 1834, X, p. 391.

[20]See William Buckland, "Notice on the *Megalosaurus* or great fossil lizard of Stonesfield," especially 391.

[21]See Hooykaas, "Geological uniformitarianism and evolution," 12–19.

[22]Constant Prévost, "Observations sur les schistes oolithiques de Stonesfield en Angleterre, dans lesquelles ont trouvés plusieurs ossemens fossiles de mammifères," see 406 and 415–416. This paper is condensed from a somewhat longer version which appeared in the *Bulletin* of the Société Philomathique.

[23]De Blainville's papers are "Doutes sur le prétendue Didelphe fossile de Stonesfield . . .," followed by "Nouveau doutes . . ." in response to Valenciennes' reply. He was supported briefly in a note by Agassiz. On the material used in de Blainville's arguments see Patsy A. Gerstner, "Vertebrate paleontology, an early nineteenth century transatlantic science," 144–147.

[24]See A. Valenciennes, "Observations sur les màchoires fossiles des couches Oolithiques de Stonesfield nommés *Didelphis prevosti* et *Didelphis bucklandi.*"

[25]See Richard Owen, "Observations on the fossils representing *Thylacotherium prevosti* (Valenciennes), with reference to the doubts of its mammalian and marsupial nature recently promulgated and on *Phascolotherium bucklandi*"; see also *Proc. Geol. Soc. Lond.* III (1838), 5–9 and 17–23.

[26]Adolphe Brongniart, *Histoire des végétaux fossiles, ou recherches botaniques et géologiques sur les végétaux renfermés dans les diverses couches du globe*, I, p. 18. On the development of paleobotany see Y. Conry, *Correspondence entre Charles Darwin et Gaston de Saporta. Precedé d'une histoire de la paléobotanique en France en XIXe siecle.*

[27]Brongniart, *Prodrome d'une histoire des végétaux fossiles . . .*, p. 1.

[28]*Ibid.*, p. 182.

[29]*Ibid.*, pp. 186–187.

[30]*Ibid.*, p. 210.

[31]*Ibid.*, p. 221. Brongniart held that the dicotelydons are the highest plants, bearing the same relationship to the vegetable kingdom as that of the mammals to the animal kingdom.

[32]Brongniart, "Considérations générales sur la nature de la végétation qui couvrait la surface de la terre aux diverses époques de formations de son écorce," see 255. This paper also appeared as a pamphlet under the same title. The translation is from the English version printed under the title "General considerations on the nature of the vegetation which covered the surface of the earth at the different epochs of the formation of its crust . . .", see 368–369.

[33]On the difference between Lamarck and Geoffroy, see Hooykaas, *Natural Law and Divine Miracle*, pp. 88–89. Geoffroy's own recognition of the difference between his "unity of plan" concept and the old chain of being is described in Isodore Geoffroy Saint Hilaire *Vie, traveaux et doctrine scientifique d'Etienne Geoffroy Saint Hilaire*, pp. 131–133.

[34]The general debate is described by Jean Piveteau, "Le débat entre Cuvier et Geoffroy Saint Hilaire"; and by Théophile Cahn, *La Vie et l'oeuvre d'Etienne Geoffroy Saint Hilaire*, pp. 194–210. The debate over paleontology and evolution is described by Franck

Bourdier, "Geoffroy Saint Hilaire versus Cuvier; the campaign for paleontological evolution (1825–1838)."

[35] See Geoffroy Saint Hilaire, "Le degré d'influence du monde ambiant pour modifier les formes animales; quéstion interessant l'origine des espèces téléosauriennes et successivement celle des animaux de l'époque actuelle," especially 80–82.

[36] See Hooykaas, *Natural Law and Divine Miracle*, p. 85.

[37] For a description of Serres' work, see E.S. Russell, *Form and Function. A contribution to the history of animal morphology*, pp. 79–83.

[38] See Geoffroy Saint Hilaire, *Philosophie anatomique; des organes respiratoires sous le rapport de la détermination et de l'identité de leur pièces osseuses*, I, pp. 448–449. Geoffroy's work on monstrosities is described in the second volume.

[39] "Le degré d'influence du monde ambiant," 83 and 86.

[40] Geoffroy Saint Hilaire, "Des recherches faites dans les carrières du calcaire oolithiques de Caen, ayant donné lieu à la découverte de plusieurs beaux échantillons et de nouvelles espèces de téléosaurus," see 57–58. Geoffroy also supposed that further changes must have been responsible for the modification of the Tertiary mammals into those of today.

[41] "Le degré d'influence du monde ambiant," 68.

[42] *Ibid.*, 80.

[43] *Ibid.*, 79–82. Geoffroy distinguished only three periods in the earth's history: the "antediluvian" era of the *Teleosaurus*, the Tertiary, and the present.

[44] See Walter Cannon, "William Buckland."

[45] William Buckland, *Vindiciae Geologicae, or the connexion of geology with religion explained*, pp. 14–15. Buckland's *Bridgewater Treatise* (discussed elsewhere) was essentially a continuation of this approach.

[46] See James Parkinson, *Organic Remains of a Former World*, I, *The vegetable kingdom*, p. 252.

[47] *Ibid.*, III, *Starfish, echini, shells, insects, amphibia and mammalia*, pp. 450–451.

[48] Parkinson's full description of these birds is given *ibid.*, pp. 302–307. For Cuvier's reaction to the same remains, see *Ossemens fossiles* (1812), III, "Mémoire sur les ossemens d'oiseaux qui se trouvent dans les carrières de pierres a plâtre des environs de Paris," especially p. 6.

[49] *Organic Remains*, III, p. 455.

[50] See William Kirby, *On the Power, Wisdom and Goodness of God as manifested in the Creation of Animals, and in their History, Habits and Instinct*, I, p. 4. This work also contained a number of strange fundamentalist speculations which were generally ignored.

[51] Smith's ideas were developed in the 1790s but did not become widely known until the publication of his map of England and Wales in 1815. For a survey of the literature on Smith, see Joan M. Eyles, "William Smith: some aspects of his life and work."

[52] Sir Everard Home, "Some account of an animal more nearly allied to the fishes than to any other class of animals," (1814), 571–577; see also *ibid.* (1816), 318–321; (1818), 24–32, and (1819), 209–216. Home tried to popularize the name *Proteosaurus*; the name

Ichthyosaurus was coined by Koenig in the *Synopsis of the Contents of the British Museum* (1818).

[53] W.D. Conybeare and H.T. De la Beche, "Notice of a new fossil animal forming a link between the *Ichthyosaurus* and the crocodile." W.D. Conybeare, "On the discovery of an almost perfect skeleton of the *Plesiosaurus.*"

[54] William Buckland, "Notice on the *Megalosaurus* or great fossil lizard of Stonesfield."

[55] Gideon A. Mantell, "Notice on the *Iguanodon*, a newly discovered fossil reptile from the sandstone of Tilgate forest in Sussex."

[56] Mantell, "The geological age of reptiles," *Edinb. New Phil. J.* (1831).

[57] For a more detailed account of the events described in this paragraph, see Sir A. Geikie, *The Founders of Geology*, pp. 410–432.

[58] A. Sedgwick and R.I. Murchison, "On the structure and relations of the deposits contained between the primary rocks and the Oolitic series in the north of Scotland", (read 1828); see the section on "Fossil fish of the Caithness strata," 141–144. For a list of the fish discovered in the Old Red Sandstone by 1835, see *Proc. Geol. Soc. Lond.* II (1835–36), No. 2, 201. Hugh Miller's *The Old Red Sandstone* of 1841 (discussed elsewhere) describes an enormous number of different forms.

[59] See John Phillips, *Figures and Descriptions of the Palaeozoic Fossils of Cornwall, Devon and West Somerset; observed in the course of the Ordnance Geological Survey of that district*, e.g., p. 161. The terms correspond to "old life," "middle life" and "new life." Sedgwick had originally used "Palaeozoic" to cover just the Cambrian and Silurian, but Phillips extended it to the Devonian and Carboniferous.

[60] Robert Jameson, *A system of Mineralogy, comprehending oryctognosie, geognosie, mineralogical chemistry, mineralogical geography and economical mineralogy*, III, p. 82.

[61] See e.g., *ibid.*, pp. 81–82.

[62] See Cuvier, *Essay on the Theory of the Earth* (1825), p. 406; and Alexander von Humboldt, "On petrifactions, or fossil organic remains," 21. This latter is borrowed from Humboldt's *A Geognostical Essay on the Superposition of Rocks*, p. 45.

[63] See *Edinb. New Phil. J.* VII (1829), 349.

[64] Robert Grant, "Observations on the nature and importance of geology," especially 297. Grant does not describe the vertebrate progression in detail. Michael Bartholomew has suggested to me in a letter that it was by reading Grant's papers that Lyell's interest in Lamarck was first aroused. On Grant's general lack of influence, however, see Chapter 4, note 85.

[65] For the quotation, see Charles Lyell, *Principles of Geology*, I, p. 145; it is derived mainly from pp. 152–153 of the edition of the *Consolations* cited below, but with the addition of a number of smaller passages.

[66] Sir Humphrey Davy, *Consolations in Travel, or the last days of a philosopher*, 5th edition, p. 143 (1st edition, 1829).

[67] W.D. Conybeare, "Letter on Lyell's *Principles of Geology*," and "An examination of those phaenomena of geology which seem to bear most directly on theoretical

speculation." See also "Report on the progress, actual state, and ulterior prospects of geological science," especially p. 407.

[68] M.J.S. Rudwick, "A critique of uniformitarian geology: a letter from W.D. Conybeare to Charles Lyell, 1841," see especially 281–282.

[69] [William Whewell], "Lyell's *Principles of Geology*, Vol. II," see 117. Whewell was referring back to the first volume. In his *Philosophy of the Inductive Sciences*, Whewell appears to abandon even directionalism in favor of pure catastrophism, arguing that Lyell has no guarantee that periods of upheaval have not alternated with periods of repose; see II, pp. 670–671.

[70] See for instance Cannon's "The problem of miracles in the 1830s," footnote 4. I have only noted the phrase "progressive development" once in the book, and then used in ambiguous circumstances; see R.I. Murchison, *The Silurian System, founded on geological researches in the counties of Salop, Hereford, Radnor, Montgomery, Caermarthen, Brecon, Pembroke, Monmouth, Gloucester, Worcester and Stafford; with descriptions of the coalfields and overlying formations*, I, p. 582.

[71] Adam Sedgwick, presidential address (1831), see 305.

[72] Sedgwick to Agassiz, April 10th, 1845; see Elizabeth C. Agassiz, *Louis Agassiz, his life and correspondence*, I, p. 384.

[73] William Buckland, *Geology and Mineralogy considered with reference to Natural Theology*, I, pp. 282–283, footnote. Buckland's main description of the cooling earth theory may be found on pp. 40–41.

[74] *Ibid.*, p. 107.

[75] *Ibid.*, p. 80.

[76] *Ibid.*, pp. 114–115.

[77] *Ibid.*, p. 294.

[78] Gillispie, *Genesis and Geology*, p. 146.

3

The New Progressionism

For the vast majority of naturalists in the 1830s, the idea of the transmutation of species was dead—their work was built firmly on the assumption that each organic form was miraculously created by the Deity. When Robert Chambers' anonymously published *Vestiges of the Natural History of Creation* revived the bogey of transmutation in 1844, they rallied round the traditional viewpoint and subjected the new hypothesis to a vitriolic series of attacks. Yet the sheer violence of their reaction has to some extent obscured the fact that a serious division had now arisen within the creationists' own camp. Workers such as Adam Sedgwick attacked *Vestiges* in the name of the original utilitarian concept of design, according to which the benevolence of the Deity was to be observed in the wonderful series of adaptations by which He fitted the species He created into the succession of environments produced by the earth's physical development. But there was now an alternative form of creationism—chiefly the work of the Swiss naturalist Louis Agassiz—in which the creative design could be seen more readily in the overall plan linking the sequence of forms into a harmonious pattern culminating in man. Agassiz objected no less violently than Sedgwick to transmutationism, yet in his recognition of a progressive plan of creation he anticipated at least one aspect of Chambers' system. Both believed that the advance of life represented the unfolding of a divine plan; they differed essentially over whether this occurred continuously or via a series of distinct but related miraculous creations.

In effect, Agassiz had introduced a totally new form of progressionism by treating the ascent of life toward man as the key to the historical process observed in the fossil record, not as a mere by-product of changing conditions. For Sedgwick, Buckland, and the followers of the utilitarian concept of design, progression was a simple consequence of the fact that the history of the earth was directional. Conditions had at first been unsuitable for the higher forms of life, but as the environment gradually came to approximate more closely that of today, more advanced creatures were successively introduced. The creation of man fulfilled the purpose of the whole system, but the fact that man occupied this high position was revealed directly—it could not have been predicted from the sequence of earlier forms. For Agassiz, on the other hand, design could also be traced out in the form of a general plan linking all of the vertebrate species and leading toward man as the perfect expression of the type. The progression of life was the gradual unfolding of this plan and had occurred, at least to some extent, independently of the changing conditions. This was indeed (to use Eiseley's phrase) a "man-centered, transcendental progressionism," since Agassiz held that it was possible for the naturalist to show that man was the key to the whole system. By thus integrating man symbolically but firmly into nature, he put forward a totally new interpretation of progressionism which stressed the existence of more or less continuous trends between the fossil forms. Such links were of limited significance in the utilitarian system, where each species was treated as a case of perfect adaptation, but they were the very essence of Agassiz's belief that nature formed a harmonious pattern.

Agassiz's was certainly not an evolutionary system in the modern sense of the word; in fact it provided a whole alternative philosophy in which "evolutionary" trends could be recognized and given a significance quite different from that which we now accept. The element of continuity in this new concept of design was totally spurious by Darwinian standards, yet the idea that there might be *intellectual* links between the species was at least a first step toward the recognition of what we would now call patterns of evolutionary development. Agassiz had produced a compromise between the hope of seeing nature as a continuously advancing system and the old philosophy of creationism in which species were absolute and distinct entities. Organic forms were really ideas in the mind of the Creator, and the fact that there were symbolic links between them confirmed the supernatural basis of the historical process because such a pattern could not have been produced by a mere physical development governed by natural law. Obviously, Agassiz would never be able to accept Darwinism, where the element of design was totally eliminated. But Chambers' system, as we are now beginning to realize, was quite different from Darwin's, and it was based on a transcendental concept of design that was paradoxically quite similar to Agassiz's. Chambers too believed that there was a preordained plan of development which ensured that progression was the key feature in the history of life. But he took what seemed to him the obvious step of suggesting that the plan was built into the universe at the creation, designed to unfold in continuous trends without the Creator's direct interference. The physical link between the species so strenuously denied by Agassiz was in fact a direct consequence of the intellectual link embodied in the original plan. As it was originally expressed, the creation of new forms in Chambers' system was still indirectly supernatural, and although this became less obvious in his later editions it was never completely repudiated. Indeed, Chambers even retained something of the old concept of species by arguing that progress takes place in small but discontinuous leaps from one form to the next. The modern reader familiar with the Darwinian kind of continuity in nature may find it difficult to appreciate the difference between a series of linked miracles and transmutation brought about by supernaturally preordained steps. But in the mid-nineteenth century this difference was crucial and ensured that someone like Agassiz could still not accept Chambers' system. Despite the element of transcendental design upon which both based their idea of development, the question of physical continuity forced Agassiz to reject *Vestiges* in the same way as Sedgwick and the followers of the old kind of design. The real problem was the origin of man: if there was a physical as well as a symbolic link between him and the animals, all sorts of philosophical consequences would follow. Chambers accepted these consequences and argued that they were not as unpleasant as everyone had assumed, but this point damned his system in the eyes of almost all his contemporaries.

Louis Agassiz
and Transcendental Progressionism

Louis Agassiz was born into a Protestant Swiss family in 1807.[1] He studied first at Heidelberg and then at Munich, where he worked with Ignatius Dollinger and acquired a lifelong interest in embryology. At Munich he also heard the lectures of Lorenz Oken and F.W.J. von Schelling, which introduced him to the vision of a wide ranging philosophy of nature. In later years Agassiz was reluctant to admit that he had absorbed the influence of *Naturphilosophie*, but it is probable that such an influence was at least partially responsible for his interest in the construction of a general plan of organic nature. Philosophers such as Oken and Schelling saw the universe as a unified system centered on man and urged that the whole could only be understood as a kind of spiritual development. Both occasionally spoke as though this had manifested itself in the historical development of life, although in general they let it be assumed that the process could only be represented in an idealized manner.[2] Certain German embryologists, however, had been rather more explicit. They proposed the "law of parallelism," which implied that the mammalian embryo passes successively through stages corresponding to the hierarchy of the vertebrate classes. Friedrich Tiedmann in 1808 and J.F. Meckel in 1821 both suggested that this development parallels that which has taken place in the history of life on the earth.[3] At first there had been only very limited evidence for this, but by 1830 the fossil record was beginning to offer more support and it is perhaps not surprising that Agassiz, with his early interest in embryology, should have sought to put this kind of progressionism on a firmer footing.

Agassiz's essential difference from the *Naturphilosophen*, however, lay in his staunch refusal to accept the speculative side of their developmental idealism. His father was a Protestant pastor, from whom he absorbed a simple Christian faith which allowed him to treat the progression of life as the unfolding of God's plan, worked out through a series of miraculous creations. His conviction that the form of each species is fixed at its creation may also have been reinforced by a brief period of study under Cuvier in 1831. Here he began the study of fossil fishes which was to make his reputation as a paleontologist, a study continued after he returned to Switzerland to take up a post as professor of natural history at the newly formed Academy of Neuchatel. The sections of his great *Recherches sur les poissons fossiles* were published throughout the decade following 1833. In 1846 he sailed for the United States to begin a whole new phase of his life and work.

Although developed in the 1830s, Agassiz's theory of progression was first described in some detail in an address delivered to the Academy of Neuchatel in 1842 and in the introduction to the collected edition of the *Poissons fossiles*.[4] The basis of the system was the advance through the vertebrate classes up to man himself. Agassiz denied that there has been an increase in the level of organization among any of the three invertebrate types.[5] Nor would he admit that there was any progression between the types, i.e., in their order of introduction. He argued that "the Radiata, Mollusca, Articulata and Vertebrata appeared simultaneously as the first inhabitants of the earth."[6] This was a rather strange position, since workers such as Sedgwick and Murchison were just beginning to discover that there were periods uninhabited by vertebrates at the beginning of the fossil record. But Agassiz clung to his original belief and was reluctant to abandon it even in his *Essay on Classification* of 1857.[7] Although others would take the appearance of the first vertebrates as a progressive step, for Agassiz the only meaningful progression was the advance within the vertebrates toward the perfection of the type in man. He was able to argue that there was now clear evidence for an Age of Fishes at the beginning of the fossil record, followed in turn by an Age of Reptiles and then of Mammals.[8] The development of life could thus be represented as a tree, with ever higher branches shooting out from the parent trunk.[9]

Agassiz felt that little effort need be wasted on demonstrating that the succession of classes in the fossil record represented the ascent of a hierarchy of complexity.[10] Such a hierarchy was, in fact, the basis of his vertebrate classification, although, especially in his later career, he admitted that it would be impossible to reduce the whole type to a linear plan.[11] There was certainly a central thread running through the type: the fishes were "the point of departure for a graduated series which commences with them and by them, to terminate in man himself."[12] It was this ascent toward man which, for Agassiz, formed the central theme of the history of life. He did attempt briefly to trace out a series of physical developments paralleling the organic ones—a gradual increase in the amount of dry land is mentioned here, and we know from his theory of glaciers that he accepted the cooling-earth hypothesis.[13] Such directional physical changes might explain why successive populations had to become extinct to make room for the next phase of the plan; Agassiz followed Cuvier in believing that a species could only survive while the conditions for which it was created still prevailed, and the concept of ice ages that he derived from the cooling-earth theory provided a convenient addition to the actual mechanisms for producing worldwide extinctions. But such changes were not presented as the real driving force behind the progression observed in the fossils. Instead, he argued that the Creator has worked with a "premeditated plan, connected together in all its parts" and that the aim of the plan was the production of man rather than the adaptation of forms to new conditions.

> The history of the earth proclaims its Creator. It tells us that the object and the term of creation is man. He is announced in nature from the first appearance of organized beings; and each important modification in the whole series of these beings is a step towards the definitive term of the development of organic life. It only remains for us to hope for a complete manifestation in our epoch of the intellectual development which is allowed to human nature.[14]

Here we see the transcendental element in Agassiz's taxonomy serving as the foundation of his progressionism. This point was to remain the basis of his teaching throughout his career—it was developed in 1845 in a lecture on the "Plan of creation" and in a number of books published in English after he had settled in America.[15] It was also expounded in private letters to workers such as Sedgwick, although we have seen that the latter preferred to avoid the transcendental approach and seek evidences of design only in the adaptation of successive creations to new conditions.[16] Sedgwick may have been reluctant openly to dispute Agassiz's new approach to the argument from design, but he almost certainly realized that it introduced a completely new picture of the history of life in which progression toward man was the central theme.

Although committed to the idea of an unfolding plan of nature, Agassiz was determined to show that the process could only have taken place discontinuously, through a series of miraculous creations. He argued that no scientist familiar with the laws of physiology could believe "that the first reptile which lived on the earth descended by means of generation, or in any other manner, from any of the fishes which existed anteriorly."[17] The fossil record itself "does not at all harmonize with the systems taught, and which formerly represented the whole of these organized beings as forming a graduated series, rising without interruption from beings the most imperfect to man, who now reigns supreme on the earth."[18] In other words, the plan of nature has been unfolded mainly through a series of sudden leaps, when new and important variations on the vertebrate theme were introduced. These leaps corresponded to the introduction of the various classes, and *within* each class there was no sign of a progression that might link it with those above and below. In his own field of study—the history of the fishes—Agassiz knew that there was little evidence for a gradual progression. He had divided the class into four orders based on the characters of the scales: the Placoids, Ganoids, Cycloids, and Ctenoids.[19] The fish of the earliest periods all belonged to the Placoid and Ganoid orders and included many forms with a cartilaginous skeleton. The Cycloids and Ctenoids, comprising most of the modern bony fishes, seemed to appear suddenly at the beginning of the Cretaceous. This could not however, be regarded as a progressive step, since as Buckland had already pointed out, some of the earlier fishes—notably the Sauroids of the Carboniferous—were far closer to the reptiles in structure than any modern fish. Even the earlier fish of the Old Red Sandstone combined cartilaginous skeletons with highly developed nervous and generative systems. They certainly could not be regarded as very primitive forms, a fact emphasized repeatedly by Hugh Miller in opposition to transmutation. Agassiz admitted privately to the worried Sedgwick that there must be some significance in the appearance of the Sauroids just before the first known reptiles.[20] But as long as the history of the class was surveyed from beginning to end there was no sign of a gradual progression. (See Plate VI.) All opponents of transmutation in the period after Chambers' *Vestiges* took this as clear evidence that there was no trend within the fishes leading toward the next higher class. They tended to ignore the rather obvious point that further changes within the fishes after the appearance of the reptiles could not be expected to have any relevance to the general ascent through the classes.

Although Agassiz started out believing that the unfolding of the plan of nature was a discontinuous process, the logic of his position seems eventually to have forced him into a re-evaluation of the evidence. He was never, of course, converted to evolution and the absolute continuity of living development, but he does seem to have come to accept that the gaps between the successive creations were not as large as he at first thought. It was quite possible for him to recognize lines of development connecting the successive forms—at one point in the *Poissons fossiles* he presented a diagram to illustrate the history of the fishes which looks almost as though it were drawn by an evolutionist.[21] For Agassiz, however, the lines indicated imaginary connections between distinct species in God's plan of creation, not a gradual evolution. But he had appreciated that there could be general trends in the history of life, and his interest in embryology led him to see that this might apply to the overall progression. As Jane Oppenheimer has pointed out, he became fascinated for a time with the so-called recapitulation theory, according to which the overall history of life and the development of the individual embryo follow out the same basic plan.[22] In the *Poissons fossiles du vieux grès rouge* he argued for such a connection within the fishes and implied that it might also apply at a wider level.

> But what I wish to prove here . . . is the truth of the law now so clearly demonstrated in the series of vertebrates, that the successive creations have undergone phases of development analogous to that of the embryo in its growth and similar to the gradations shown by the present creation in the ascending series, which it presents as a whole. One may consider it as henceforth proved that the embryo of the fish during its development, the class of fishes as it at present exists in its numerous families, and the type of fish in its planetary history, exhibit analogous phases through which we may follow the same creative thought like a guiding thread in the study of the connection.[23]

In other words, the taxonomist discovers a hierarchy of organization within the class, which has also served as the plan of development both for the class as a whole and the individual. In his *Twelve Lectures on Comparative Embryology* of 1846, Agassiz explicitly applied the law of parallelism to the whole sequence of vertebrate classes and again implied that this was also the plan to be discovered in the fossil record.[24] Since embryological growth is obviously continuous, his desire to see the same plan here as in the overall development of life inevitably led him to accept that the latter process was more continuous. He always claimed that there were enough discontinuities in the fossil record to disprove transmutation, but in some later works he admitted that if we take an overview of the record it gives the impression of a general progression.[25] His concept of an overall plan of creation had thus led Agassiz to de-emphasize, at least to some extent, the discontinuities in the advance through the classes that were often used as primary arguments against transmutation.

It is obvious that Agassiz had to imagine quite a complex overall plan of vertebrate development: there might be a central theme running through toward man, but the later expansion of the fishes was to some extent a separate phenomenon governed by its own logic. But the system was fairly flexible in this respect. Since species were distinct units or entities linked only abstractly in the mind of God, it was possible to imagine that the central plan had side branches representing variations on the main stages of development. There was no general progressive trend that must force all forms along the same preordained path—in this way Agassiz avoided a difficulty that Chambers was never able to deal with properly. The problem was that the more complex the fossil record revealed itself to be, the more the transcendentalist had to stretch his imagination and his faith to retain the idea that there really was a unifying theme running through the whole process. In later works such as the *Essay on Classification*, Agassiz's description of the history of life had to introduce a number of new concepts in order to deal with the complex relationships now being discovered: there were embryonic types, prophetic types, and progressive types, all representing the different kinds of trends that could now be distinguished within the sequence of forms. Agassiz, at least, was still able to retain his belief that the whole bewildering array still rested upon a vast, unifying plan in the mind of the Creator. But in a later chapter we shall see that in the course of the 1850s a number of naturalists began to feel that the fossil record was now too complex and disorganized for it to be treated as the product of a harmonious program of development.

Some of these later naturalists agreed that the fossil record exhibited fairly continuous trends linking the succession of forms (although they did not necessarily accept transmutation). But in the 1840s there were very few who were prepared to follow Agassiz in admitting that the sequence revealed any degree of continuity. The vision of life unfolding according to a harmonious divine plan of creation did not go unnoticed, but even those who were converted to the new approach continued to share the common belief that the record confirmed the complete discontinuity of the advance. The clearest example here is Hugh Miller, the Scots stonemason turned geologist who was encouraged in his early career by Agassiz. In his *Old Red Sandstone* of 1841, Miller supported the new approach by discussing the recapitulation theory. Quoting Serres on the law of parallelism, he argued that the same pattern of development could be seen in both embryology and the fossil record, a sure indication that the history of life was governed by an overall progressive plan.[26] But Miller never accepted that the advance was continuous—we shall see below that much of the *Old Red Sandstone* was devoted to the fossil evidence showing that there was no progression within the classes and hence that each must have been introduced suddenly. He made it very clear that his main fear was the transmutation theories of earlier writers such as Lamarck. To trace out a relatively continuous process within the succession of fossil populations was to give the transmutationist just the evidence he needed, so the right-thinking naturalist ought to do his best to show just how large the gaps in the record really were. In fact, Miller had anticipated correctly. Chambers' *Vestiges* appeared only three years later, arguing for the continuous unfolding of the plan of creation through transmutation and using as one of its main lines of evidence precisely the interpretation of the fossil record that Miller had been at such pains to refute.

Robert Chambers: Progression and Transmutation

Robert Chambers was certainly not a professional scientist: he *few were* started life as a bookseller and eventually joined his brother in establishing a prosperous Edinburgh publishing house. But he was in contact with a number of naturalists in the Edinburgh circle and was passionately interested in scientific studies. He felt that the professionals tended to be over-cautious in their approach and around 1836 conceived the project of preparing a universal synthesis of scientific knowledge. Convinced that the universe is ruled completely by laws imposed on it by the Deity, he was naturally led to reject the miraculous creation of species. But as M.J.S. Hodge has recently pointed out, it would be a mistake to picture Chambers as merely an unsuccessful forerunner of Darwin.[27] In fact, the system he eventually presented anonymously in his *Vestiges of the Natural History of Creation* in 1844 was built on totally different principles. The most important difference in the present context is that to some extent Chambers' rather unusual concept of natural law included what Agassiz had meant by a "plan of nature." In effect he tended to see the development of life as a preordained or designed process, but since this was supposed to be worked out via a law rather than a mere abstract plan, it had to take place continuously. The element of design became less obvious in later editions of the book, but its striking presence in the first edition shows how important it was for the formulation of Chambers' ideas.

Chambers' main problem had been to find a law which could be used to synthesize the known facts about the history of life. As Hodge points out, it was embryology which gave him the clue, allowing him to see that the whole process could be seen as a continuous progressive development.[28] But once having decided on this basis, there were a number of different ways in which it could be applied. He could have adapted the old progressionism to his ends by claiming that the advance of life had been produced in response to changing physical conditions. Chambers' geology was not catastrophist, but it *was* directional: he accepted the nebular hypothesis of the formation of the solar system and mentioned several commonly accepted directional theories of the earth in connection with individual stages in the history of life.[29] Hooykaas has actually suggested that he accepted the physical changes as the cause of the living development, more or less after the fashion of Geoffroy Saint Hilaire.[30] But Hooykaas admits that in later editions Chambers repudiated this approach, and in fact the first edition is equally difficult to interpret in this way. It is clear that Chambers' concept of law allowed him to believe that spontaneous generation and transmutation could be triggered off automatically whenever certain conditions were manifested, but he always maintained that the nature of the new forms produced in each case was predetermined by the Creator. The physical development of the earth may thus have limited the rate at which the progression took place, but the real cause of the later appearance of higher forms had to be sought elsewhere. Nor did Chambers' explanation anticipate that which was later proposed by Herbert Spencer, who saw progression as an inevitable trend underlying the constant efforts of life to adapt to new conditions. For Spencer, there was only a general tendency toward higher levels of organization, with the actual path of development being decided by the conditions themselves. But Chambers made it clear that the whole course to be followed in the evolution of life was built into the universe from the beginning.

The fact that Chambers was interested in the idea of a transcendental plan of organic structure is evident from the chapter he devoted in his first edition to the circular or quinary system of taxonomy proposed by William Sharp MacLeay.[31] This strange system—which interested a number of contemporary naturalists— arranged all taxonomic ranks into units of five. The animal kingdom was divided into five types, each of which in turn could be divided into five classes and so on (Cuvier's Radiata were divided into two types and the amphibians accepted as a distinct vertebrate class). It was a highly artificial system, admirably suited for giving the impression that the whole organic creation has been formed according to a rational— even an artistic—plan. The second clue to Chambers' early thinking is the discussion of natural law introduced into his first edition. He quoted extensively from Charles Babbage's unofficial *Ninth Bridgewater Treatise*, a work which had proposed a highly unusual concept of law in an effort to show that even miracles could in a sense be regarded as phenomena produced by laws.[32] Babbage used the analogy of his calculating engine (a forerunner of the computer) which, he pointed out, could be programmed to read off a series of numbers and then at a predetermined point switch to a new series. To an observer looking at the dials of the machine, the change would appear to be a totally inexplicable event—a "miracle"—but actually it would have been programmed into the hidden workings from the beginning. Chambers used this notion of a "higher" law programmed occasionally to interfere with the normal laws of nature to explain how one species might suddenly give rise to another. By implication, every step in the advance of life must have been designed from the first by the Great Programmer in accordance with His master plan.

MacLeay's quinary system was not a very good one to use as the basis of a theory of linear progression, although it did give Chambers a way of showing how man stood at the head of the animal kingdom.[33] In general he simply accepted that there was a more or less linear hierarchy among the vertebrates and that this has represented the historical plan of ascent.[34] Since it was embryology which had first suggested this to him, he was able to use the law of parallelism as the basis of his general theory of organic development. He implied that the basic plan of both embryological and racial development is predetermined; normally, an embryo simply develops along the scale until it reaches the point corresponding to its full level of organization and then branches off from the main plan to complete the minor details of its structure. All that would be necessary for a new and higher species to be formed would be for the embryo to carry on a little further up the scale before the process was completed. A slight extension of the period of gestation to allow this extra development to take place was all that was needed.[35] Although at one point he hinted that changing conditions might produce this extension, Chambers' main explanation of the phenomenon was in terms of Babbage's "higher" law interfering with the normal process of generation. Thus

> . . . the simplest and most primitive type, under a law to which that of like production is subordinate, gave birth to the next type above it . . . this again produced the next higher, and so on to the very highest, the stages of advance being in all cases very small—namely, from one species to another; so that the phenomenon has alway been of a simple and modest character.[36]

The progressive plan of creation is gradually revealed as forms are produced having passed further and further along the predetermined scale of embryological development. The law governing the succession of forms in the history of life incorporates the plan of creation in the form of the pattern of embryological development, the whole having been programmed into the universe from the beginning.

To the extent that Babbage's higher law represented only a series of disguised miracles, there is an obvious similarity between Chambers' system and that of Agassiz. The fact that Chambers often referred respectfully to Agassiz's work[37] may indicate that he at least recognized some affinity between their ideas, however strongly Agassiz himself would have repudiated this. But *Vestiges* probably represents an independent discovery of the idea that embryology is a guide to the general plan of development—certainly the book's manner of presentation differed widely from anything that had gone before. From the first edition onward there were passages which suggested a far more naturalistic theory of transmutation than would be implied by Babbage's concept of law. Thus, for instance, Chambers argued that the spontaneous generation of life could still occur whenever the right conditions were produced, as in the (soon discredited) experiments of

Crosse and Weekes on the production of insects by electricity.[38] The element of design could be retained by arguing that the new form was preordained to appear whenever these particular conditions were realized, but the overall implication was that life could be produced simply through chemical and electrical activity. The later editions also maintained that the central tendency toward progress could be modified by the external conditions to give the adaptations praised by the natural theologian. To many this all sounded horribly materialistic and seemed to deny the main point of Babbage's argument that each case of the higher law's operation was individually predesigned. Although Chambers retained his quotation from Babbage in some of his later editions, the general trend of his modifications always made the theory appear more materialistic. The discussion of MacLeay's highly artificial system of classification, for instance, was soon abandoned. Nevertheless, Chambers certainly proposed his system as an explanation of universal development based on design—he was never an atheist, whatever his opponents might say about him. He may have gradually abandoned this mode of description as he tried to make his theory more scientifically respectable, but we shall see that certain implications of it always remained.

Approximately one third of *Vestiges* was devoted to the fossil evidence which had now become available to suggest that the history of life really was a progression paralleling the development of the vertebrate embryo. Essentially what Chambers wanted to do was to show that a progression could be traced out within each class, so that it started with its lowest members (i.e., those closest in structure to the previous class) and ended with the highest. This was by no means an easy task in 1844, since only the basic sequence of the classes could readily be shown to fit the pattern. Since the theory implied that very simple forms of life had first been spontaneously generated at an early point in the earth's history, Chambers was able to point to the invertebrates of the Cambrian and Silurian as evidence that lower forms had indeed come first. The appearance of the higher vertebrate type in the form of the fishes of the late Silurian and Old Red Sandstone clearly represented a progressive step.[39] In an attempt to show that there has been a progression within the first vertebrate class, he argued that the early forms belonged to "the order of cartilaginous fishes, an order of mean organization and ferocious habits, of which the shark and the sturgeon are living specimens."[40] To back up this highly debatable opinion of the first fish he went on to discuss the better known forms of the Old Red Sandstone in more detail. He noted

that the external armor of forms such as *Cephalaspis* indicated at least
a superficial resemblance to the invertebrates[41] and stressed that the
early fish resembled the embryonic structure of the modern members
of the class in that they possessed a heterocercal or unevenly branched
tail.[42] (See Plate VII.) This last was quite a useful point, since even
opponents of transmutation such as Hugh Miller admitted it to be
true, although they often tried to conceal its significance by arguing
that in other respects these fish were far from primitive. Chambers did
not hesitate to emphasize that the Sauroid fish of the Carboniferous
indicated that the class had reached its most reptilian form immediately before the first true reptiles appeared.[43] The reptiles themselves,
however, were very difficult to fit into the picture of continuous
progression, since there was no sign at that time that the amphibians
had appeared before the higher members of the class. Chambers could
do little more than state the facts as they were known and pass rapidly
on to the mammals.[44] Here there was rather more scope, since the
Oolitic marsupials were universally admitted to belong to the lowest
order of the class.[45] Some forms of development had clearly taken
place during the Tertiary, although many workers denied that this was
a true progression and the only indisputably progressive step was the
final appearance of man.[46] Chambers had done his best to make the fossil
record conform to his theory, but in a number of places the relationship
was either highly debatable or one of total incompatibility.

The first editions of *Vestiges* elicited so many refutations that
Chambers was forced to make extensive revisions and to publish a
sequel which he entitled *Explanations*. In particular, he had to defend
himself against the critique written for the *Edinburgh Review* by Adam
Sedgwick, who had taken upon himself the task of explaining just how
discontinuous the fossil record really was. Sedgwick emphasized that
in many respects the first fish had quite a high level of organization,
and Chambers now conceded that they could not be dismissed as
primitive just because they lacked a fully ossified vertebral column.[47]
But he pointed instead to the fact that their muscles must have been
attached to their external "skeleton"—a point "strikingly indicating
their low grade as vertebrate animals."[48] Concerning the reptiles,
new evidence was introduced: the discovery by A.T. King of fossil
footprints in the Carboniferous rocks of Pennsylvania.[49] Since it had
been reported that these prints were probably made by batrachians
(amphibians), it was now possible to argue not only that the reptiles
had been introduced earlier that had hitherto been supposed but also
that they had first appeared in their lowest form. Sedgwick had

claimed that the Oolitic marsupials had no connection with the main development of the mammals in the Tertiary since the intervening Cretaceous rocks revealed no sign of mammalian remains. Chambers countered this by arguing that the Cretaceous rocks were all laid down in the depths of the sea where no mammalian bones could ever be expected to have found their way.[50] He also tried to support his belief that the marsupials could have descended from a lower form by citing the work of Thomas Rymer Jones, professor of comparative anatomy at King's College, London.[51] Jones had suggested that in some ways the marsupials formed almost a link between the birds and the true mammals. The only problem here was that there was no sign of an "Age of Birds" linking the reptiles with the mammals, so the anatomical resemblance had little real relevance to the line of development suggested by the fossil record.

As well as defending himself against specific attacks, Chambers also tried to create a more sophisticated basis for his theory. Many of his improvements tended to blur the original purity of his vision in which the development of life had been presented as the unfolding of a more or less linear plan. Yet traces of his earlier way of thinking remained, and this is especially true of his attempt to deal with the increasing complexity of the fossil record. It was rapidly becoming obvious that the attempt to treat the development of life as a purely linear process was hopelessly unrealistic because of the sheer number of different fossil forms being discovered in the various strata. Even Agassiz, although he continued to feel that there was a key thread running through the whole process up to man, nevertheless realized that the growth of each class would have to be treated as though it were multilinear. The system which was eventually worked out represented the history of each class as a number of lines diverging away from a common point of origin, each in the direction of some particular state of adaptation. Chambers too realized that

> . . . development has not proceeded, as is usually assumed, upon a single line which would require all the orders of animals to be placed one after the other, but *in a plurality of lines in which the orders and even minuter subdivisions of each class, are ranged side by side.*[52]

But the explanation he gave of this phenomenon differed widely from that eventually worked out by the paleontologists and clearly illustrates that the concept of a divinely preordained plan still influenced his thinking. He introduced the term *stirpes* to describe the separate lines of development, declaring that the affinities between species would have to be sought "vertically" along such lines rather than "horizontally" between the present members of each taxonomic group. A number of distinct stirpes existed parallel to one another, and each could pass separately across the boundaries that the taxonomist recognizes between the classes.[53] In other words the orders making up a class today are *not* related by community of descent from a single ancestral form—the resemblance between them is caused by the fact that every line of development is programmed to pass along the same hierarchy of complexity. Each of the stirpes has certain distinguishing characteristics which define the order, but the basic features of the class are fixed by the common plan along which they all must pass. Thus Chambers retained to some extent the linearity of his original thinking about the ascent of life, along with the essential belief that the progression through the classes is designed according to a predetermined plan.

Chambers suggested that within each class every line must originate with a marine form of life, since "Fluid, required for all embryonic conditions, is also necessary to the origination of the various stirpes of both kingdoms."[54] Each line passes through forms adapted to different habitats on the order: sea water, fresh water, shores, marshes, jungles, dry elevated plains, and finally mountains. But this is not purely a sequence of adaptations—Chambers explicitly connected it with a hierarchy of organization so that the real driving force could still be seen as the progressive trend.[55] In the fourth edition of *Vestiges* he proposed a number of very strange family trees which help to illustrate just how far his thinking was from that which we accept today. The mammals of the Tertiary, for instance, must all have arisen from marine forms, although Chambers never showed how these might be connected with earlier reptilian species. He explained the huge bulk of many pachyderms by suggesting that they were derived from herbivorous cetacea such as the manatee and dugong.[56] Another line ran from the seals throught the otters and polecats up to the dog family.[57] At this point he seems to have forgotten about the various forms of the early Tertiary which bear no close resemblance to any modern species. He had become carried away by the prospect of arranging the modern mammals into a series of linear patterns that would fit his preconceived ideas of a planned progression. Nevertheless, the new system he presented did appear at least superficially more sophisticated, and the idea of parallel lines of development allowed him to explain the somewhat embarrassing fact that, despite the essentially progressive plan of creation, there were still many of the lower forms around. Evidently the rate of progression must vary, so that there will be some stirpes which have not yet climbed all the way up the scale of organization.[58]

The most controversial aspect of *Vestiges* was its wholehearted acceptance of man as the last step in the continuous vertebrate progression. For Agassiz, man stood only symbolically at the head of creation—he was not actually connected to the lower forms and his physical perfection served only to emphasize the significance of his quite distinct moral and intellectual faculties. Chambers, however, accepted that man was just the highest animal, whose mind has evolved along with his body. By accepting the phrenologists' claim that the brain is the organ of the mind he was able not only to explain how the two developed together but also to illustrate that the mind itself must be governed by law. Man in fact, was no exception to the universal reign of law. This might at first seem degrading, he admitted, but once this view was accepted it enabled one to see all the more clearly that the central purpose for which the whole system of nature was designed must be the eventual production and refinement of intelligence. He was even prepared to hint that the process might continue in the future, and man himself be superseded by some yet higher form in which this quality was even more pronounced.

It was these opinions on man which damned Chambers' book in the eyes of most of his contemporaries. Naturalists who were prepared to condone, if not to accept, Agassiz's claim that man was the high point of vertebrate creation were deeply shocked as soon as the link with the animal kingdom was made into a real and not a symbolic one. There was an immediate public outcry against *Vestiges*, in which scientists from all backgrounds vied with one another in condemning it. To a certain extent this was an easy job. Especially in his earlier editions, Chambers had introduced a number of very debatable scientific points that only confirmed his amateur status. From the beginning his concept of "law" —including as it did an element of design or guidance—was extremely unorthodox and alienated even a scientist such as T.H. Huxley, later one of Darwin's chief supporters. In fact, Chambers pointed the way toward the whole shadowy area of late nineteenth century theistic evolutionism, where both theologians and scientists tried to blur the distinction between law and design in order to adapt the new philosophy of continuous development to the old religion. But this was at a later stage, and to begin with there was general opposition to the basic idea of transmutation itself, in which paleontology formed the basis of perhaps the most consistently used argument.

The apparent discontinuity of the fossil record in the 1840s cannot be ignored as a factor influencing the debate over Chambers' book. Indeed, it seems probable that paleontology played a crucial role at this period in both a positive and a negative sense. It can hardly be a coincidence that Agassiz and Chambers expounded their two quite different interpretations of the new progressionism in the 1840s, when the evidence for the full sequence through the vertebrate classes had just been completed. The idea that progression was the central theme in the history of life had been suggested before, especially by embryologists, but it had remained merely a speculation until the fossil record itself forced naturalists to take it seriously. In previous decades it had been far easier to interpret the smaller number of progressive steps merely as evidence in favor of the prevailing directional theories of the earth, but now the sheer extent of the advance made it difficult to fit the whole process into such a scheme. The alternative interpretation was already waiting to be combined with the fossil evidence to form the new progressionism of Agassiz and Chambers. But at the same time the sequence of fossils was still very fragmentary and irregular. Agassiz's approach was relatively easy to equate with a certain degree of discontinuity in the advance, but Chambers had been forced to ignore or reinterpret a good deal of the detailed evidence. Clearly the opposition to *Vestiges* drew its strength from the horror with which many viewed the book's implications for man, but to present the opponents as men determined to twist the evidence in order to demolish a theory they hated is quite unrealistic. Just how fragmentary the evidence for progression appeared at the time is the the subject of the following chapter.

Notes

[1] For an account of Agassiz's life and work see the biography by Edward Lurie, *Louis Agassiz. A life in science.* Lurie's views on Agassiz and evolution are summed up in his "Louis Agassiz and the idea of evolution." See also Ernst Mayr, "Agassiz, Darwin and evolution."

[2] For a brief discussion of Oken's ideas see Russell, *Form and Function*, p. 215. See also the Ray Society's translation of Oken's *Lehrbuch der Naturphilosophie* under the title *Elements of Physicophilosophy* (London, 1847), especially the introductory passages where the ambiguity of the evolutionary process is very pronounced. A somewhat more realistic interpretation is implied on pp. 185–186, where Oken suggests that life

originated in the sea. On Schelling and evolution, see for instance Fr. F. Coppleston, *A History of Philosophy*, VII part 1, p. 141. A modern reader turning to Schelling's *The Ages of the World*, trans. F. Bolman, Jr., expecting to find a realistic discussion of the development of life will soon realize that this is purely an idealistic reconstruction of the Absolute's self-realization in the universe and in consciousness.

[3] For a discussion of Tiedmann's *Zoologie* and Meckel's *Beyträge zur vergleichenden Anatomie* see Russell, *Form and Function*, p. 73 and pp. 90–94. K.E. von Baer also spoke of the fossil evidence for progression in 1822, but the embryology he expounded in his *ÜberEntwickelungsgeschichte der Thiere* (1828) demolished the law of parallelism between the development of the embryo and the hierarchy of the classes; see Boris E. Raikov, *Karl Ernst von Baer, 1792–1876, sein Leben und sein Werk*, pp. 64–69. On these developments in general see Owsei Temkin, "German concepts of ontogeny and history around 1800."

[4] Agassiz's address was published in French, but I have only had access to the English translation: "On the succession and development of organized beings at the surface of the terrestrial globe; being a discours delivered at the inauguration of the Academy of Neuchatel," which appeared in the *Edinb. New Phil. J.* in 1842. See also Louis Agassiz, *Recherches sur les poissons fossiles . . .* , I, introduction. This combined edition was actually published in 1844; for the dates of the publication of individual parts see A.S. Woodward and C.D. Sherborn, *A Catalogue of British Fossil Vertebrata*, pp. xxv. Agassiz had set out his ideas in a much shorter introduction issued in 1833.

[5] Agassiz, "The development of organized beings," 396. He later changed his opinion on this point; see for instance his *Essay on Classification*, ed. Edward Lurie, (originally the first volume of the *Contributions to the Natural History of the United States*, 1857), p. 106. Note that from the beginning Agassiz was prepared to admit a progression in the size of the invertebrates, although this is very different to a progression in complexity; see his *Monographie d'échinoderms, vivans et fossiles*, part 2 "Des Scutelles," p. 22.

[6] "The development of organized beings," 392.

[7] *Essay on Classification*, p. 26.

[8] "The development of organized beings," 393.

[9] *Ibid.*, 397.

[10] *Ibid.*, 395.

[11] *Essay on Classification*, pp. 32–33.

[12] "The development of organized beings," 395.

[13] See *ibid.*, 394. On Agassiz's glacier theory and its importance for his view of the history of life, see M.J.S. Rudwick's essay review of the *Studies on Glaciers*, translated by Albert V. Carozzi.

[14] "The development of organized beings," 399. For a later but very explicit statement that progression is the result of God's plan and has nothing to do with changing physical conditions, see *Essay on Classification*, pp. 103–104.

[15] This lecture is discussed in Elizabeth C. Agassiz, *Louis Agassiz, his life and correspondence*, I, p. 396. Later works in which Agassiz developed this theme include the *Principles of Zoology* he produced in 1848 in collaboration with A.A. Gould, the *Essay on Classification* of 1857, and the *Structure of Animal Life* of 1862.

[16]See Agassiz's letter to Sedgwick, June 1845; E.C. Agassiz, *Louis Agassiz*, I, p. 392. Sedgwick's reaction was discussed in the previous chapter.

[17]"The development of organized beings," 391.

[18]*Ibid.*, 390–391.

[19]See Agassiz, "On a new classification of fishes, and on the geological distribution of fossil fishes," and the introduction to the *Poissons fossiles.*

[20]Agassiz to Sedgwick, June 1845; E.C. Agassiz, *Louis Agassiz*, I, 393.

[21]See *Poissons fossiles*, I, facing p. 170.

[22]See Jane Oppenheimer, "An embryological enigma in the *Origin of Species.*"

[23]Louis Agassiz, *Monographie des poissons fossiles du vieux grès rouge, ou système devonien*, p. 24. The translation is from E.C. Agassiz, *Louis Agassiz*, I, pp. 369–370.

[24]Louis Agassiz, *Twelve Lectures on Comparative Embryology, delivered before the Lowell Institute in Boston, December and January, 1848–49*, pp. 26–27 and 96–98. To defend the law of parallelism, Agassiz explicitly attacked K. E. von Baer's embryology as it had been described in English by Martin Barry.

[25]See for instance *Essay on Classification*, pp. 108–109, and Louis Agassiz and A.A. Gould, *Principles of Zoology; touching the structure, development, distribution and natural arrangements of the races of animals, living and extinct*, p. 227 and pp. 237–238. Note, however, that by this time Agassiz had accepted von Baer's embryology and was no longer using the concept of recapitulation except *within* the classes.

[26]See Hugh Miller, *The Old Red Sandstone, or new walks in an old field*, pp. 241–243.

[27]See M.J.S. Hodge, "The universal gestation of nature: Chambers' *Vestiges* and *Explanations.*" On Chambers' life, see Milton Millhauser, *Just before Darwin. Robert Chambers and Vestiges.*

[28]Hodge refers to the explanatory preface Chambers wrote for the 10th (1853) edition of *Vestiges*; see "The universal gestation of nature," 136–139.

[29]For example, Chambers noted that there was no dry land when the first living forms appeared and that the appearance of the higher vertebrates corresponded with a decline in the carbon dioxide content of the air; see [Robert Chambers], *Vestiges of the Natural History of Creation* (London, 1844, reprinted by the University of Leicester Press, 1969), pp. 56, 73, 89 and 150–151. The edition of *Vestiges* cited here has an introduction by Sir Gavin de Beer.

[30]Hooykaas, "The parallel between the history of the earth and the history of the aminal world," 13.

[31]*Vestiges*, pp. 236–276. MacLeay's system had first been proposed in his *Horae Entomologicae* of 1819–21. He later went to live in Australia.

[32]*Vestiges*, pp. 206–210. Babbage himself was interested in the application of his idea to the production of new species, and quoted part of the correspondence between Charles Lyell and Sir J.F.W. Herschel on this topic; see *The Ninth Bridgewater Treatise, a fragment*, pp.226–227.

[33]*Vestiges*, pp. 262–273.

[34]See, for instance, *ibid.*, pp. 191–192. Chambers here gives the vertebrate hierarchy as reptiles, fish, birds, and mammals, but this is presumably a misprint—he had no intention of upsetting the normal arrangement.

[35]*Ibid.*, p. 213.

[36]*Ibid.*, p. 222.

[37]T. H. Huxley described Chambers' constant references to Agassiz as evidence of just how out of date the science in *Vestiges* was! See his review of the 10th (1853) edition, 338.

[38]*Vestiges*, pp. 185–190.

[39]*Ibid.*, p. 58.

[40]*Ibid.*, p. 63.

[41]*Ibid.*, p. 69.

[42]*Ibid.*, p. 71.

[43]*Ibid.*, p. 90.

[44]*Ibid.*, pp. 95–99.

[45]*Ibid.*, p. 112.

[46]*Ibid.*, p. 114.

[47][Robert Chambers], *Explanations: a sequel to the Vestiges of the Natural History of Creation*, pp. 50–54.

[48]*Ibid.*, p. 52.

[49]*Ibid.*, p. 57. For the original report, see A.T. King, "Footprints in the coal rocks of Westmoreland Co., Pennsylvania." Reptilian footprints had been discovered in the Carboniferous as early as 1841 by W.E. Logan, Director of the Geological Survey of Canada, but they had been incompletely reported and were ignored; see Bernard J. Harrington, *Life of Si William E. Logan*, pp. 115–116.

[50]*Explanation*, pp. 91–92.

[51]*Ibid.*, p. 89. See Thomas Rymer Jones, *General Outline of the Organisation of the Animal Kingdom and manual of comparative anatomy*, p. 760. Chambers would have used the first edition of 1841.

[52]*Explanations*, p. 69. Chambers' italics.

[53]*Ibid.*, p. 73.

[54]*Ibid.*, p. 70.

[55]*Ibid.*, p. 71.

[56]*Vestiges*, 4th edition (London, 1845), p. 266. The Pachydermata are the "thick skinned" animals of Cuvier, including forms such as the elephant.

[57]*Ibid.*, pp. 270–271.

[58]See, for instance, *ibid.*, pp. 259–260.

4

The Opponents of Progression

The previous two chapters have shown how the fossil evidence gradually emerged for a number of progressive steps in the history of life. We have seen how the British and French naturalists at first found it easier to interpret these steps as evidence for the directionalist theories of the earth which they favored. The idea of a specifically progressive trend in creation was discussed in Germany, but only became a serious issue in the rest of Europe when Agassiz showed that the new fossil evidence could be used to support such an interpretation. Finally Chambers had pushed this to its logical conclusion by suggesting that the progressive trend was built into the universe and programmed to unfold of its own accord. But it has been noted that the evidence for the succession of the vertebrate classes was still very irregular, so that Chambers at first found it by no means easy to fit into the picture he was presenting. There were many naturalists, in fact, who found the irregularities so great that they questioned the whole idea of progression. The most obvious case is that of Charles Lyell, who supported his uniformitarian world view by maintaining a complete "anti-progressionism." He argued that even the ascent through the classes was illusory, caused by the fragmentary nature of the fossil record. Others admitted the basic sequence through the classes but held that within each class there was no sign of a progression. These supporters of what can be termed "discontinuous progression" denied that there was a general trend affecting the whole history of life—changes occurring within the classes were not related to the more fundamental step-by-step ascent.

In discussing the opposition of many scientists to the progressive evolutionism of writers like Chambers, some historians have argued that from the beginning it should have been obvious to an unbiased mind that the underlying pattern in the fossil record was one of continuous development.[1] A.O. Lovejoy, for instance, implied that once the basic sequence through the classes had emerged, paleontologists ought immediately to have realized that any irregularities were the result of imperfection in the available record. It is true that some of the more important gaps were soon filled in; but even in 1859 Darwin had to devote a whole chapter of the *Origin of Species* to explaining away those that still remained. In the 1840s the "imperfection of the geological record" was far worse—it was not just that there were gaps between the known forms, but the actual sequence in which the major groups appeared was completely irregular. There were, in fact, two quite separate problems, each of which was enough to throw serious doubt on the idea of continuous progression. First, as the century progressed the origin of each class was gradually pushed further back down the geological time scale as new evidence came to light. This was compounded by a number of misidentifications, but the problem was obvious enough for writers like Lyell to argue that there was no guarantee that further discoveries might not extend the history of even the higher classes back to the beginning. The second and more obvious problem was the level of organization of the first known examples of each class. All too often these were not the lowest members of the class, so there was no obvious progression within the group. Lyell argued that this was incompatible with any theory of progression and good evidence of the inadequacy of the known record as a true guide to the history of life. The supporters of discontinuous progression, however, took the same information at its face value, accepting the basic ascent through the classes but emphasizing that there was nothing to indicate the existence of a progressive trend within the classes that could link one with another.

It cannot be denied that the opponents of continuous progression were inspired by other considerations that a pure desire to analyze the available evidence. Lyell had both theoretical and emotional reasons for rejecting the advance of life, and those who attacked *Vestiges* on the basis of discontinuous progression shared his intense dislike for any system that demeaned the status of man's mental and moral faculties by incorporating him into the animal kingdom. Of course these considerations affected the naturalists' judgment of the fossil record. This is most obvious in the case of the early fish, which often combined both high and low features in the same structure: Chambers stressed the low features and his opponents naturally concentrated on the high ones. But in some cases the evidence against progression was unequivocal and had to be accepted by both sides of the debate. Indeed there were Continental naturalists who did not at all share the theological concerns of their British counterparts, but who were nevertheless just as convinced that the theory of continuous progression was mistaken. This indicates quite clearly just how bold a step Chambers had taken beyond what was indicated by the available fossil record.

Charles Lyell
and Anti-Progressionism

Although Charles Lyell was trained for the law, he acquired an early interest in geology and absorbed the catastrophist doctrines which prevailed in the mid-1820s. Gradually, however, he came to the conclusion that it was unscientific to postulate that past conditions were different from those of today. He participated in the Geological Society's debates which eventually discredited the supposed reality of the Biblical deluge. In 1830 he launched his major attack on catastrophism with the publication of the first volume of *Principles of Geology*, in which his aim was "to explain the former changes of the earth's surface by reference to causes now in operation."[2] This commitment to actualism was, in fact, derived from his revival of James Hutton's strict uniformitarian vision of an earth which has remained more or less in a steady state throughout the period open to the investigations of the geologist. In particular Lyell attacked the most important of the directional theories—that of the cooling earth— and in the third edition he proposed an alternative explanation of the earth's central heat.[3] Inevitably his program also had to include a rejection of the prevailing belief that the late introduction of the higher organic classes was evidence of directionally changing physical conditions. But he also attacked what was becoming the basis of Agassiz's new progressionism—the belief that there has been a progressive trend in the history of life independent of the external conditions. Indeed, his was the first attempt to make a clear distinction between what we have called the old progressionism and the new.

Lyell began his discussion of the fossil record by quoting from Sir Humphrey Davy's *Consolations in Travel,* which, he claimed, presented the picture of a "progressive development of life from the simplest to the most complicated forms."[4] The composite quotation that he gave tends to stress the very slight emphasis on a process of development "preparatory to the existence of man" in Davy's work at the expense of the explanation based on changing physical conditions. Lyell then went on to accuse Davy and many other naturalists (whose names he did not specify) of supporting an inconsistent position which combined two concepts of progression that should really be treated as alternatives. They had endeavored "to connect the phenomena of the earliest vegetation with the nascent condition of organic life, and at the same time to deduce from the numerical predominance of certain types of form, the greater heat of the ancient climate."[5] The mention of a vegetable progression here makes it possible that Lyell had Brongniart's theory in mind, although neither Davy nor the French writer had really stressed the idea of an independent progressive trend. Lyell went on to insist that the fossil record cannot be used simultaneously to support *both* the theory of a specifically progressive tendency in creation *and* the belief that the earth has gradually been cooling down.

> If . . . the prevalence of particular families be declared to depend on a certain order of precedence in the introduction of different classes into the earth, and if it be maintained that the standard of organization was raised successively, we must then ascribe the numerical preponderance in the earlier ages of plants of simpler structure, *not to the heat*, but to those different laws which regulate organic life in newly created worlds.[6]

This point is quite valid: the same evidence cannot be used to support the two alternative positions. If Brongniart, Davy, and others wished to explain the development of life as a result of their directional theories of the earth, they could not at the same time hint at an independent progressive trend in nature. They could recognize a *de facto* progression but they could not treat this as a causal factor governing the history of life.

After thus distinguishing between the two possible explanations of progression, Lyell then went on to argue that in fact the fossil record did not provide sound evidence for the supposed advance of life. He began by attempting to dispose of the claim that the early populations revealed the earth's temperature to have originally been much higher. He agreed that the primitive plants of the Carboniferous period were suited to a higher temperature, but pointed out that the same condition of the fossil record would have been produced if the earth's land masses had then been positioned closer to the equator than they are today.[7] There could have been more or less cyclic changes in the nature of successive populations produced by the fact that the continual elevation and subsidence of the land would at times create continents around the poles, at other times in the equatorial regions. The apparent progression since the Carboniferous was thus illusory, and the cooling-earth theory was not the only possible explanation of the observations.

Turning to the animal kingdom, Lyell tried to show that there has been no absolute increase in the level of living organization. He argued that the higher forms, the mammals for instance, may always have existed, but during some periods formed only a very small proportion of the total population. The changing proportions could once again be explained as a result of the fluctuating conditions produced by the changing locations of the continents. He explained the absence of mammalian remains from the Secondary rocks by combining this idea with what he supposed to be the great imperfection of the fossil record. The remains of mammals were, he pointed out, the least likely to find their way to the sea bed where new rocks are being laid down. Hence there will be fewer mammalian fossils than those of any other class, and it is hardly surprising that none have been found from periods when the class was reduced to comparative rarity by adverse conditions. There was in any case one stratum of Secondary rock which had revealed a few mammalian remains—the Oolitic slate of Stonesfield, containing the marsupials noted above in Chapter 2. Lyell mentioned these fossils, along with the possibility that the bones of cetacea had also been found in Secondary rocks.[8] In later editions of the *Principles* he described the Oolitic marsupials in considerable detail. His main point was that here in the center (not at the end) of the so-called Age of Reptiles was a small number of mammals. The Oolitic *did* reveal the presence of the highest class, the later Cretaceous rocks did *not*; the mammals only reappeared in the Tertiary, by which time they had become the dominant forms of vertebrate life.

> On comparing, therefore, the contents of these [Cretaceous] strata with those of our oolitic series, we find the supposed order of precedence inverted. In the more ancient system of rocks, mammalia, both of the land and the sea, have been recognized, whereas in the newer, if negative evidence is to be our criterion, nature has made a retrograde, instead of a progressive movement, and no animals more exalted in the scale of organization than reptiles are discoverable.[9]

In other words, the latter part of the Secondary appears to show a degradation rather than a progression. But Lyell's real intention was to show that we cannot base our interpretation of the history of life on "negative evidence." Mammals are found in the Oolitic, so almost certainly they existed both before and after that period in numbers so small that their remains have not been found. Hence, concluded Lyell, there is no real evidence for the so-called advance of life represented by the apparently late appearance of the mammals.

It is obvious why Lyell should take so much trouble to reject the claim that progression gives evidence of directional changes in the earth itself. But it is not quite so clear why he should have been equally concerned about the possibility of a distinct progressive trend in creation, especially since in 1830 this idea was still largely confined to the more speculative German naturalists. We know, however, that Lyell himself had toyed with the concept of a fundamentally progressive universe during the early years when he was still a catastrophist. In a review written in 1826 he commented on the apparent progression in the fossil record and quoted Butler's *Analogy of Religion* to the effect that man finds himself in the midst of a progressive scheme of things.[10] When he developed his uniformitarian system he soon appears to have realized that this idea now represented a potential threat, although not in quite the same way as the alternative explanation based on changing conditions. The chapter on progression in the *Principles* reveals why he could no longer accept a specifically progressive trend: he now saw that it could all too easily imply that man should be treated merely as the last and highest animal. He asked quite bluntly if the human species was to be "considered as one step in a progressive system by which, as some suppose, the organic world advanced slowly from a more simple to a more perfect state?"[11] He had already in effect denied this by rejecting progression, yet he still found it necessary to insist that man could not in any case be regarded as the highest animal. He argued that "the superiority of man depends not on those faculties which we share in common with the inferior animals, but on his reason, by which he is distinguished from them."[12] To treat man as the highest animal was to "strain analogy beyond all reasonable bounds."[13] Agassiz's transcendental taxonomy would, of course, connect man with the animals in just this way, even though Agassiz would have agreed that in his moral and intellectual faculties he was quite distinct. Lyell, however, was so concerned on this issue that he was now prepared to reject even the claim that in his physical structure man represented the peak of the vertebrate hierarchy. His real aim was to establish a complete discontinuity between man and the animals, and the denial of progression was merely a second line of support for this position. By separating man from a totally unprogressive animal kingdom, he hoped to make a bridging of the gap impossible even for those who might disagree with him over the taxonomic relationship between them.

Agassiz was able to keep his system within the bounds of orthodoxy because his belief in miraculous creations allowed him to treat man as morally distinct from the animals despite the physical resemblances. Lyell's distrust of progression arose because his support for the principle of continuity in nature led him to prefer some form of continuous species production to miracles. Not that he contemplated anything that we should now recognize as evolution; but in a well-known letter to Sir J.F.W. Herschel in 1836 he made it clear that he thought the "origination of new species" to be the work of "intermediate causes."[14] He still felt that each new species would have to be a product of the Creator's benevolent design and foresight, and a reference to Babbage's work in this and another letter of the same period suggests that Lyell, like Chambers, was interested in the possibility of a higher law built into the universe and programmed to create new species at regular intervals in accordance with a preordained plan.[15] Without specifically accepting that the higher law would work via transmutation, Lyell nevertheless would have seen that such a system might appear to connect man with the rest of the animals, especially if there was a progressive trend among the lower forms leading toward him. He both sensed and feared the implications that Chambers was to draw out of the combination of continuity and progression, and his outright rejection of the latter was a necessary precaution required by his campaign to establish the reign of continuity. Michael Bartholomew has recently summed up the evidence suggesting that it was his fear for the status of man that kept Lyell from accepting a full-fledged transmutation theory throughout most of his career.[16] The uniformitarian world view was not, as has sometimes been alleged, anti-evolutionary: it was anti-developmental, but the rejection of catastrophic extinction and the emphasis on continuity in the history of the earth led naturally to the assumption that the introduction of new species was also a continuous process. In Lyell's case continuity obviously cannot be equated with transmutation, but the threat that man might be incorporated into the rather vague concept of continuous creation was enough to make him reject progressive development, so that the first appearance of the moral faculties in man would have to be seen as a distinct event—the one discontinuity that he was willing, even anxious to accept.

The paleontological discoveries of the 1830s made Lyell's position less secure by extending the scope of the apparent progression, but he refused to accept that the evidence had become conclusive. Successive editions of the *Principles* continued the attack on progression, and in 1851 he took the opportunity to present his views in a presidential address to the Geological Society. He still made it clear that one of his main concerns was to show that man was not "the last term in a regular series of organic development" by undermining the evidence for the developments themselves.[17] He began once again with the vegetable progression, which Brongniart had reaffirmed in his *Tableaux des genres des végétaux fossiles* of 1849. According to Lyell, the earliest fossil plants were not all of a low level of organization, and in any case they came only from marine deposits which could not be expected to tell us much about the terrestrial vegetation of the time.[18] He concentrated, however, on the supposed vertebrate progression, developing a number of points which can, in fact, be reduced to two basic arguments. On the one hand he stressed the latest discoveries which had revealed that the classes actually appeared much earlier than geologists of previous decades had anticipated, and on the other he emphasized the complex structures of many of these early forms, a point which he felt to be inconsistent with the claim that the overall sequence was one of progression.

Within the animal kingdom, Lyell noted that the earliest representatives of all three invertebrate types are often of a high level, and that the first fish in the Silurian included cestraciont sharks "than which no ichthyic type is more elevated."[19] Again, we have no evidence of the terrestrial populations of the earlier periods, although we know that dry land existed; the land *could* have been inhabited, since the latest dredging operations had confirmed Lyell's original claim that terrestrial forms very rarely find their way down to the deep sea bed, where most of the early rocks were laid down.[20] Although it had long been thought that the reptiles did not appear until after the Carboniferous, he was able to point to a number of recent discoveries which revealed that the class had existed when the coal beds were laid down. There were numerous reptilian footprints, for the validity of some of which Lyell himself could vouch, and more recently the actual remains of reptiles had been discovered in the Carboniferous rocks of Germany.[21] In addition he claimed that some of the later Secondary reptiles—*Ichthyosaurus*, for example—were considerably closer to the warm-blooded forms that those of today.[22] Numerous footprints of birds had now been discovered in the Trias, again showing that a class had existed much earlier than had originally been supposed. Lyell was here referring to the footprints discovered by Edward B. Hitchcock in the New Red Sandstone of the Connecticut valley. These are now thought to have been made by dinosaurs, but at this time most authorities attributed them to birds.[23] Concluding with the mammals, Lyell noted that Richard Owen was now beginning to suspect that the Oolitic "marsupials" might actually be true placentals.[24] The mammals of the Tertiary showed no sign of a progression toward the human form, confirming that the final appearance of man would have to be treated as a complete discontinuity.[25]

Arguments based on the high status of the first members of each class were not likely to impress Lyell's contemporaries, since most of them accepted this and incorporated it into the discontinuous version of progression. But the tendency for new discoveries to push each class further back in time was just what his position would predict, offering the prospect that eventually even the highest forms might be found at the beginning of the fossil record. A further discovery along these lines was made just in time for Lyell to add a note on it to the end of his 1851 address. A series of footprints was found in the Silurian rocks of Canada by William E. Logan, the director of the Canadian Geological Survey, and were at first identified by Owen as the tracks of a freshwater tortoise.[26] Had this interpretation been maintained it would have placed the origin of the reptiles far earlier than anyone but Lyell would have believed possible. Further discoveries were made in the following year, but Owen now changed his mind and claimed that the marks had actually been made by some form of invertebrate.[27] This destroyed what had appeared to be a striking vindication of Lyell's position, but another almost equally spectacular discovery soon took its place. Gideon Mantell identified the skeleton of a reptile in rocks which were thought to belong to the period of the Old Red Sandstone, naming it *Telerpeton elginense.*[28] Roderick Murchison—now emerging as a strong supporter of progression—grumbled privately about Lyell's open jubilation at this new indication of the early origin of the reptiles.[29] Eventually Murchison had his revenge, when he reported in a note prefaced to the fourth edition of his *Siluria* that T.H. Huxley had identified other characteristic forms showing that the stratum from which *Telerpeton* was derived belonged to the New, not the Old, Red Sandstone.[30] But this was not until 1867, by which time even Lyell had abandoned his opposition to progression for entirely different reasons.

Lyell's original position also gained support from new evidence concerning the mammalian population of the Secondary era. As early as 1847 a fossil mammal had been discovered in the rocks of the upper Triassic, showing that the class had been in existence at least one period earlier than had been supposed.[31] As Lyell noted in 1851, opinion was also changing on the subject of the Oolitic mammals from Stonesfield: *Amphitherium*, at least, was now regarded as a placental insectivore rather than a marsupial. In 1854 another placental, *Stereognathus ooliticus*, was described, again showing that the population of this period was not limited to the "lower" marsupials. In the same year a mammalian fossil was discovered in the Purbeck beds of the Wealden formation (immediately above the Oolitic), and Owen suspected that this, too, was by no means a primitive member of the class.[32] Two years later, as a result of direct encouragement by Lyell, the Purbeck beds were explored in detail by an amateur geologist named S.H. Beckles, with the result that an impressive range of mammalian remains was brought to light.[33] Lyell was extremely enthusiastic about these developments, since they all helped to substantiate his claim that it was only the imperfection of the fossil record which had prevented naturalists from seeing that the Secondary era might have been populated throughout by a small number of mammals.

Lyell maintained his opposition to progression throughout the 1850s. His recently published notebooks from this period show that he was still grappling with the problem of the origin of species and was still convinced that the phenomenon was "supernatural," even though it might be continuous and hence in conformity with law. Natural selection as it was being formulated by Darwin was an inadequate cause and would have to be supplemented by some power which actually formed new varieties and species.[34] This was to remain his opinion even after he nerved himself to accept evolution in the 1860s. He gave a noncommittal account of Darwin's theory in his *Antiquity of Man* in 1863 and only adopted a more positive tone in the tenth edition of the *Principles*, five years later. But he made it clear that he was still disturbed about the status of man,[35] and in the end it was progression which showed him a way out of this dilemma. After serving so long as his main bulwark against the connection of man with the animals, it was now to become the vehicle through which he finally accepted the link. In the *Antiquity of Man* he admitted that he now accepted progression as a "useful" or even an "indispensable" hypothesis.[36] But he found it strange that the Darwinians in fact had less interest in the possibility of a progressive trend in creation than many of their opponents. He was now coming to believe that the gradual advance of life represented the direct control of the Deity over evolution. A.R. Wallace had suggested that God has guided the last stages of human evolution, and in a letter to Darwin in 1869 Lyell hailed this view and extended it in effect to the whole range of the advance of life. He believed

> ... that the Supreme Intelligence might possibly direct variation in a way analogous to that in which even the limited powers of man might guide it in selection, as in the case of the breeder or horticulturalist. In other words, as I feel that progressive development or evolution cannot be entirely explained by natural selection, I rather hail Wallace's suggestion

that there may be a Supreme Will and Power which may not abdicate its functions of interference, but may guide the forces and laws of nature.[37]

Thus the connection of man with the rest of the animal development would become acceptable provided it was agreed that the whole upward trend of life revealed God's direct role in the production of higher faculties, including the moral ones that finally crowned the process.

Although Lyell eventually abandoned his opposition to progression, it must not be forgotten that for three decades he was able to find enough anomalies in the fossil record to make his original position seem not implausible. There were other naturalists too who shared his distrust of the evidence for progression and directionalism. In Britain these included Gideon A. Mantell and, at a later date, Edward Forbes. Mantell had made important contributions toward establishing the Age of Reptiles, but this did not prevent him from accepting uniformitarianism and the consequent anti-progressionism of Lyell's position. In his popular *Wonders of Geology* he noted the existence of the Stonesfield mammals in the middle of the Age of Reptiles and insisted that "some of the fossil animals which first appear in the strata belong to families with a highly developed organization," exactly the reverse of what the progressionist would predict.[38] Edward Forbes was the paleontologist of the Geological Survey of Great Britain, and by the early 1850s his researches had convinced him that "The scale of the first appearance of groups of any degree is most clearly not a progressive one."[39] He proposed instead his theory of "polarity" in the rate of creation of new species, based on a negation of the whole concept of directionalism in nature as complete as that of uniformitarianism itself. Since the organic world was to be seen as a manifestation of the Divine Idea, time could be treated as merely a form of representation, not an integral part of things.[40] Forbes took the middle of the fossil record as a focus from which a trend could be traced in both directions, running forward in time into the "Neozoic" era and *backward* into the Paleozoic. The two extremes of the fossil record were the periods of the "maximum development of generic types." Few of Forbes' contemporaries could make much sense out of this system, but they respected his experience as a paleontologist and his rejection of progression may not have gone unnoticed.

There were also Continental paleontologists who gave indirect support to Lyell's anti-progressionism. One of these was Constant Prévost, whose earlier support for Cuvier's interpretation of the fossil record had led him to challenge the location of the Oolitic mammals.

Now in 1845 he published a paper in which he argued that there is in fact no evidence that living forms have become more perfect in the course of time, either as the result of changing conditions or as the gradual unfolding of a hierarchical plan of creation.[41] The differences between the successive populations, he argued, were no more than would be expected from the different geographical locations in which the various strata had been laid down.[42] In a later paper Prévost supported the cooling-earth theory but claimed that this should not lead us to expect significant developments in the history of life.[43] He believed that all living forms could be arranged to form a continuum, something like the old chain of being, with the extinct forms filling in the gaps between the living ones. But he insisted that "those called *ancient* or *extinct*, and those called *new*, cannot be placed one at the head and one at the lower end of the series."[44] In other words, even though there was a hierarchical plan of creation in the taxonomic sense, this had not, as Agassiz claimed, served as the basis of the historical sequence of creations.

More systematic opposition came from the noted French stratigraphist Alcide Dessalines D'Orbigny, who was engaged in a project to subdivide the Jurassic and Cretaceous systems into a series of distinct formations. Each of these divisions was to be uniquely characterized by its invertebrate fossils—D'Orbigny was an extreme catastrophist who maintained that a worldwide upheaval annihilated all living forms at the end of each period of deposition.[45] By 1847 he had concluded that many of the invertebrate sequences he could trace out were incompatible with any theory of progression,[46] and in 1850 he published a paper extending this to all forms of life. Taking each of the four *embranchements* in turn, he attempted to refute the theory of progression by showing that many of the earliest forms were of the highest level of organization. In each of the three invertebrate types, the successive forms have either remained at the same level of organization or have actually declined in the more recent periods.[47] The order of appearance of the vertebrate classes did appear to be progressive, he admitted, but within each class there was a clear sign of a contradictory tendency.[48] The first Ganoid and Placoid fishes were among the most perfect of their orders, and the later Cycloids and Ctenoids were "less perfect than the fish of the paleozoic formation."[49] The same was true for the other classes. In another paper published in the same year, D'Orbigny supported his anti-progressionist argument by surveying the relative numbers of species within the various orders of animals.[50] He pointed out that if the general trend of life were toward progression, the primitive orders of each class should gradually decline in their total number of species, while the more advanced orders should show a corresponding increase. In many cases the fossil record actually showed the reverse of this trend—it was the more primitive species that became more numerous in later periods. The whole idea of a steady advance in the level of organization was thus an illusion.

Discontinuous Progression

what is meant by progression?

Even D'Orbigny admitted that the successive appearance of the vertebrate classes appeared to constitute a progression, and most paleontologists continued to accept this point. Distrustful of Lyell's complete anti-progressionism, they took the high status of the early members of some classes as evidence that the advance of life has been a step-by-step process. Each class was supposed to have appeared as a sudden discontinuity, and once created would continue to stay at the same level of organization or even decline. Since there was no general progressive trend to link them, each of the classes could only have been formed by the direct, miraculous intervention of the Creator. Of course, the opponents of transmutation felt that all species have been introduced by miracles, but discontinuous progression made this far easier to prove in the case of the first species in each class. By using the evidence in this way the threat of organic continuity implied by Lyell's system could be removed. The basic progression through the classes could be interpreted either in terms of the new or the old progression-ism, i.e., as the result of a progressive plan of creation or of changing physical conditions, and the introduction of man was still the last and greatest of the progressive leaps. Discontinuous progression thus became one of the leading arguments used against Chambers' transmutation theory, its points repeated over and over again by naturalists on both sides of the Atlantic who found the implications of a gradual advance toward man unacceptable. But the same arguments were also repeated by workers who were later to become strong supporters of continuous development and transmutation, a fact which clearly illustrates just how little support the fossil record of the 1840s gave to Chambers' position.

Even before *Vestiges* appeared the implications that might arise from a combination of the new progressionism with the principle of continuity had been anticipated, and steps taken to undermine any hope of such a combination being worked out. The man who did this was Hugh Miller, the Scots stonemason who turned himself into a popular writer and a geologist of some repute. Miller began his scientific work by studying the Old Red Sandstone of his native Scotland and the fossil fishes that were prominent in these rocks. His papers on this topic were collected together to form his *Old Red Sandstone* of 1841. For an expert opinion on the fish he had naturally turned to Agassiz, from whom he also seems to have absorbed the idea that the progression of life represented the unfolding of a divine plan. His interest in what we have called the new progressionism is evident from his discussion of the recapitulation theory as evidence that a unified plan governs all phases of organic development.[51] On one point, however, he was rather more cautious than Agassiz: he refused to emphasize the human form as the head of the physical hierarchy and hence as the logical goal of the progression. This was almost certainly the result of his immediate fears for the status of man, should any attempt ever be made to base a theory of transmutation on the new evidence for progression. In order to counter this possibility he was even more determined than Agassiz to demonstrate the extreme discontinuity of the fossil record, using this to show that the plan of creation must have been worked out through a series of discontinuous and obviously miraculous steps.

Since Miller was not quite so concerned with the transcendental vertebrate taxonomy developed by Agassiz, he had no reason to take up the rather artificial position that all four animal types must have appeared together at the beginning of the history of life. He argued that there was a definite progression from the annelids of the Cambrian to the crustaceans of the early Silurian and then to the first vertebrates—the fish of the late Silurian. His main intention was to demonstrate that these early fish, and also the more numerous forms of the Old Red Sandstone itself, were so high that there was no possibility of tracing out a progression within the class that could link it with the invertebrates below and the reptiles above. Ignoring Agassiz's more detailed classification, he distinguished two broad groups of fishes, those with osseous skeletons and those with cartilaginous. Neither of the two, he emphasized, could be regarded as absolutely superior to the other.

> The arrangement of the two groups is parallel, not consecutive; but the parallelism, if I may so express myself, seems to be that of a longer with a shorter line;—the cartilaginous fishes, though much less numerous in their orders and families than the other, stretch farther along the scale in opposite directions, at once rising higher and sinking lower than the osseous fishes.[52]

In the taxonomic sense there was almost complete continuity, since the lowest cartilaginous fishes were little better than worms, while the superior members of the tribe—Agassiz's Sauroid fishes—were almost connecting links with the reptiles.[53] Thus in principle there could have been an absolutely continuous progression in the history of the class, but the fossil record revealed that this was not, in fact, what had happened. Agassiz himself admitted that the Sauroids only appeared just before the first known reptiles, but Miller sidestepped this point by claiming that the very earliest fish were more or less equivalent to the Sauroids in rank. They were "cartilaginous fishes of the higher order."[54] Many of the Old Red Sandstone fishes had cartilaginous skeletons, with extensive external plating or armor, but Miller noted that many had at least a bony skull. Their high rank was derived mainly from their brains and nervous systems, which appeared to have been well-developed when compared with other and later members of the class. Since most naturalists admitted that the nervous system was of great importance, it seemed reasonable to claim that on the whole these early forms could be given a high rank. But such a claim was disputable, since—as Chambers soon pointed out—Miller's criterion was not the only one which could be used. The fact that the main part of the skeleton was cartilaginous could itself be taken as a sign of low rank, especially as the spine was supposed to be the distinguishing feature of the vertebrates. Furthermore, the relationship of the Sauroids to the reptiles was based on actual similarities in some parts of their structure, not on a vague assessment of their position on an abstract scale of organization. The various criteria for establishing the hierarchical scale were not consistent, but Miller was ignoring this in his attempt to demonstrate the lack of any real progression within the fish. It must be pointed out, however, that he was quite right to emphasize the gap between the first known fish and the invertebrates. Even today we have no fossil evidence of forms linking the vertebrates with any earlier type.

Having stated his fundamental assumption about the rank of the first fish, Miller could now quite easily claim that there has been no progression within the class. In the first edition of his book he argued for a progressive increase in the size of the fishes, but a note was added to later editions pointing out that new evidence had destroyed even this limited kind of advance.[55] With respect to the level of organization there was no progress at all. The fish remained at the same high level over several geological periods, so that the later introduction of the reptiles could only be regarded as a sudden and miraculous leap, not as the result of transmutation.

Now it is a geological fact, that it is fish of the higher orders that appear first on the stage, and that they are found to occupy exactly the same level during the vast period represented by five succeeding formations. There is no progression. If fish rose into reptiles, it must have been by sudden transformation,—it must have been as if a man who had stood still for half a lifetime should bestir himself all at once, and take seven leagues at a stride. There is no getting rid of miracle in the case—there is no alternative between creation and metamorphosis. The infidel substitutes progression for Deity; Geology robs him of his god.[56]

This viewpoint was repeated, after the publication of *Vestiges*, in Miller's *Footprints of the Creator* of 1847. He now explained that he did not find the non-miraculous unfolding of the plan of creation distasteful in itself, but as a Christian he could not accept it because of its implications for man.[57] To this end he reinforced his claims about the high status of the early fish—the book was subtitled *The Asterolepis of Stromness*, a reference to yet another highly organized fish from the Old Red Sandstone. Miller now insisted that just because some of the early fish had structures that were "embryonic" compared to those of today, this did not mean that they were lowly organized.[58] Evidently Chambers had taught him to be careful with the law of parallelism. To make the discontinuous interpretation of the advance of life seem all the more striking, he now argued for what he called a "progress of degradation" within each class. Not only were the first members supposed to be the most highly organized, but there was also a decline toward the modern period.

The general advance in creation has been incalculably great. The lower divisions of the vertebrata precede the higher; —the fish preceded the reptile, the reptile preceded the bird, and the bird preceded the mammiferous quadruped. And yet, is there one of these great divisions in which, in at least some prominent feature, the present, through this mysterious element of degradation, is not inferior to the past?[59]

Man, of course, was the last great step, but the gap between him and the animals was a spiritual one, possessing "all the breadth of eternity to come, . . . an *infinitely great* difference."[60] Nevertheless, there was still an analogy here with the "progress of degradation."

The belief which is perhaps of all others most fundamentally essential to the revealed scheme of salvation, is the belief that 'God created man upright,' and that man, instead of proceeding onward and upward from this high and fair beginning, to a yet higher and fairer standing in the scale of creation, sank, and became morally lost and degraded.[61]

In other words the degradation of the vertebrate classes was the physical counterpart of the fall of man. Miller thus provided the theory of discontinuous progression with its ultimate theological justification.

Miller's obviously sincere concern for the traditional Christian view of the fall marks him off from the majority of nineteenth century naturalists who preferred the more abstract God of natural theology. Even those who opposed transmutation themselves were upset at his explicit connection between the fossil record and revelation, probably remembering the fiasco of the "deluge" in the 1820s. But they did support his general attack on the continuity of living development, especially after *Vestiges* appeared. Perhaps Chambers' bitterest opponent was Adam Sedgwick, who prepared a critique for the *Edinburgh Review* that ran to eighty-five devastating pages. Here he presented the arguments for discontinuous progression so forcefully that Chambers was able to accuse him of ignoring even the overall advance through the classes that almost everyone except Lyell accepted.[62] In fact, as we have seen, Sedgwick did not deny progression, but he made it clear to Agassiz at this time that he interpreted it only in terms of changing conditions.[63] Unlike Miller, Sedgwick was unwilling to approve of the transcendental explanation of the advance of life even in its discontinuous form. On the discontinuities themselves, however, he was quite prepared to adopt the same position as Miller. He repeated the claim that the earliest fishes were among the highest in the class, so that there was no possibility of a subsequent progress toward the reptiles.[64] Among the reptiles themselves there was equally no sign of an advance: the first known forms were not amphibians, with their close resemblance to the fish, although such degraded types did occur later in the record.[65] The Oolitic "marsupials" could not be taken as indication of a mammalian progression, since they were separated from the main development of the class by the whole space of the Cretaceous.[66] Nor was there any obvious way in which the modern mammals were more advanced than those of the early Tertiary. In fact, Sedgwick believed that each class ultimately declined, following a "law of the rise, progress and decline of the organic families."[67]

Sedgwick continued his attack in the massive preface added to the fifth edition of his *Discourse on the Studies of the University of Cambridge* in 1851. Now he did at least admit the basic progression through the classes, quoting his own remarks from his presidential address to the Geological Society twenty years before. He also included some comments about a "gradual evolution of creative power" and "Nature's true historical progress" that sound almost as though he was now prepared to lend some support to the new progressionism.[68] He still made it clear, though, that he felt the advance toward the modern forms to be correlated mainly with directional changes in the climate.[69] He repeated his original opinions on the high status of the first members of each class, again stressing that the lack of subsequent progress was a sign that the first forms must have been miraculously created. He still argued that the Oolitic mammals could not be connected with those of the Tertiary, ignoring Chambers' counterclaim that the intervening Cretaceous rocks could not be expected to reveal mammalian fossils because they were laid down in the depths of the sea.[70] To confirm his belief that the Tertiary mammals did not in themselves reveal a progression, he quoted a letter he had received from Dr. Hugh Falconer in India. Having unearthed an impressive series of new mammalian forms from the Sewalic (or Sevalic) foothills of the Himalayas, Falconer insisted that in this sequence of Tertiary deposits the species "shew no gradation of organic rank correspondent with their successive appearance in time. The oldest are not the lowest in type—nor are those of the newest strata of a more noble degree."[71] Here, argued Sedgwick, was impressive confirmation of perhaps the most significant absence of internal progression: the mammals did not show a gradual advance that could in any way be interpreted as a preparation for the appearance of man.

Sedgwick and Miller were the two leading opponents of Chambers' book, but almost every British and American periodical—both scientific and popular—carried its own review.[72] These were almost always critical, and often made the discontinuity of the fossil record one of their main lines of attack. *Fraser's Magazine*, for instance, carried an article under the explicit title "Geology versus development" which went further than most in hoping that Agassiz's original position would ultimately be verified by the discovery of vertebrates in even the earliest rocks.[73] Nor were Miller's the only popular geological texts to urge the discontinuity of progression. Others included the *Ancient World* of D.T. Ansted, professor at King's College, London, and G.F. Richardson's *Introduction to Geology*.[74] As late as 1860 John Phillips—who had established the three great eras of the history of life—urged in his *Life on the Earth* that Sedgwick's *Discourse* still offered a complete refutation of continuous progression.[75] No doubt much of this barrage of criticism was prompted by fears for *Vestiges'* broader implications, although Chambers' eccentric choice of evidence and strange philosophy of science also alienated many professionals. Criticism on this last score is prominent in T.H. Huxley's account of the tenth edition written for the *Medico-Chirurgical Review* in 1854, which even Huxley later regretted on account of his "needless savagery."[76] Yet he also devoted a good deal of space to the fossil evidence against continuous progression, especially the high level of the early invertebrates and the first fish.[77] Since Huxley was later to become one of Darwin's most forceful supporters, it is evident that the discontinuity of the fossil record has to be taken seriously. The arguments for discontinuous progression were not concocted purely by workers whose judgment had been unbalanced by their fears for the status of man. But perhaps the best evidence for this comes from the attitude of Richard Owen, who after 1850 became a strong supporter of continuous development and even transmutation (despite his refusal to accept natural selection). During the 1840s even Owen was so impressed by the high level of many early forms that he spoke out for it unequivocally and was cited as an authority by many of Chambers' opponents.

In 1840 Richard Owen was emerging as one of Britain's leading comparative anatomists. He had been appointed Hunterian Professor at the Royal College of Surgeons in 1836, a post he retained for twenty years until he became superintendent of natural history at the British Museum. Although never directly involved with geology, Owen's skill as an anatomist made him the final court of appeal for the identification of fossils, as in the case of the Stonesfield mammals. In 1840 and 1841 he presented a "Report on British fossil reptiles" to the British Association for the Advancement of Science. Part one of this dealt with the already familiar marine forms, but in the second part Owen moved on to the terrestrial reptiles and introduced a major innovation in their classification. He proposed an order to be named Dinosauria (from the Greek for "terrible lizard"), including Buckland's *Megalosaurus* and Mantell's *Iguanodon* and *Hyleosaurus*. In fact, there were six other genera already described which would eventually be included, forerunners of the great range of reptiles that would within a few decades be seen to dominate the late Triassic, the Jurassic, and Cretaceous periods.[78] The order would be divided into two by H.G. Seeley in 1887–1888, but in the popular imagination the dinosaurs have remained the symbols of the Age of Reptiles. Yet when the order was first created in 1841 there were only nine known forms that were, or were to be, included; not much on which to base the reconstruction of a whole episode in the history of life. Indeed, there was surprisingly little evidence concerning the Age of Reptiles as a whole, and it can readily be understood that in such circumstancs even the broad outlines of the development of the class were obscured. At this time there was simply no sign at all that the true reptiles were preceded by the more fish-like amphibians (which were still regarded as merely the lowest order of the same class). The earliest positively identified reptiles were *Thecodontosaurus* and *Palaeosaurus* discovered in the New Red Sandstone of England in 1836. Owen regarded them as members of the order Lacertilia or lizards—by no means the lowest order of reptiles.[79] Eventually they were recognized as dinosaurs, but this would only raise their status, since Owen held that his new order was the closest in structure to the mammals. The reptiles thus appeared to begin with the higher forms—lizards or dinosaurs—and only in modern times did the lower amphibians appear. Owen was thus forced to admit that

> The evidence acquired by the researches which are detailed in the body of this Report, permit of no other conclusion than that the different species of Reptiles were suddenly introduced upon the earth's surface, although it demonstrates a certain systematic regularity in the order of their appearance. Upon the whole, they make a progressive approach to the organization of the existing species, yet not by an uninterrupted succession of approximating steps. Neither is the progression one of ascent, for the Reptiles have not begun by the perrenibranchiate type of organization, by which, at the present day, they most closely approach the fishes; nor have they terminated at the opposite extreme, viz., at the Dinosaurian order, where we know that the Reptilian type of structure made the nearest approach to the mammals.[80]

Although there was no overall progression through the class, it could have been argued that there was an initial advance toward the dinosaurs, followed by a decline to the present day. But Owen noted that even this could not be substantiated, since the earliest mammals (the Stonesfield "marsupials") came into existence along with, not immediately after, the first forms he recognized as dinosaurs.[81]

Owen's "Report" was very useful to Sedgwick and other writers for whom it represented exactly the kind of information they needed to use against Chambers. Nor was this the only way in which Owen sided with the supporters of discontinuous progression. In the volume of his *Lectures on Comparative Anatomy* devoted to the fishes, published in 1846, he proposed a new classification which helped to back up Miller's opinions on the high status of the first members of the class. Ignoring Agassiz's four orders, he proposed an arrangement of eleven orders and six sub-orders which could be placed on an ascending scale.[82] The sharks came at the top, and this order included the cestraciont sharks of the Silurian, the first known fish. In fact, Owen went out of his way to emphasize the high level of all the early cartilaginous fishes. He asked on what basis the cartilaginous skeleton could be regarded as an absolutely primitive sign—surely it was in fact an adaptation which was necessary for the way of life to which sharks and other predators were designed.[83] Nor could the external plating of many early forms be regarded as a sign of a low status. Here Owen specifically attacked a suggestion in Robert Grant's *Outline of Comparative Anatomy* of 1841 that this armor was a relic of the shells of the invertebrates from which these early fish had evolved.[84] Grant was one of the few open transmutationists at this time, although he soon fell into obscurity and is now forgotten except for Darwin's brief reference to his having been startled by Grant's support for Lamarck in conversations at Edinburgh.[85] The suggestion about the early fish was clearly derived from the transmutation theory, and it is rather surprising that Grant's opinion was not cited by Chambers, who made exactly the same point in his *Explanations*. But Owen had now thrown his not inconsiderable prestige behind the assessment of the first fish presented by Miller and against the possibility of any direct connection with the invertebrates. For the time being his influence could again be added to that of Chambers' opponents.

Although his work was used to support the discontinuity of progression, Owen himself refused requests from Whewell, Sedgwick, and Murchison that he should write a critical review of *Vestiges*. Instead he wrote the author a quite complimentary letter, and by 1849 he was himself arguing openly for the continuity of organic development.[86] For certain theoretical reasons which will be discussed below, Owen was strongly drawn to the idea of gradual progression, yet in the early 1840s he had to admit that the fossil evidence was inconsistent with this. There was no sign at all of a reptilian progression, nor was it possible to treat the earliest fishes as low forms related to the invertebrates. In these two areas the opinions of Sedgwick and Miller were quite reasonable, since Chambers had gone far beyond anything that was consistent with the fossil record as it was then known. Lyell's complete anti-progressionism gained only a few supporters, but hardly any reputable paleontologist could imagine that the record would ever become compatible with transmutation. The intensity of the feeling against *Vestiges* was no doubt derived from the fears that many felt over the status of man, but the paleontological arguments were supported by workers like Owen and Huxley, who were soon to become supporters of transmutation, and by Continental workers such as D'Orbigny, who did not share the theological interests of their British and American counterparts. The fossil evidence for the basic progression through the vertbrate classes was now fairly definitely established, providing Agassiz and Chambers with the essential basis they needed for the new progressionism. But it needed an enthusiast such as Chambers for the concept of development according to natural law to see the possibility that the enormous irregularities of the record might eventually be filled in. Only around 1850 did this process of filling in the gaps begin, when new evidence at last made a progression among the reptiles seem more probable. But at the same time the ever-increasing number of fossil species began to throw doubts on the whole idea that the progression of life could be measured by a more or less linear scale. It became evident that even if the process of development was continuous, it was also multilinear, and that many of the lines had no reference at all to the kind of steady progress envisaged by Chambers.

Notes

[1] See Gillispie, *Genesis and Geology*, p. 175; Millhouser, *Just before Darwin*, p. 97 and p. 178; A.O. Lovejoy, "The argument for organic evolution before the *Origin of Species*, 1830–1858," especially pp. 378–391. To be fair, all of these writers do admit that the fossil record was discontinuous, but they somehow manage to imply that it was only transmutationists like Chambers who really saw the significance of the patterns it revealed.

[2] Charles Lyell, *Principles of Geology, being an attempt to explain the former changes of the earth's surface by reference to causes now in operation.* On the development of Lyell's thought, see Leonard G. Wilson, "The intellectual background in Charles Lyell's *Principles of Geology*, 1830–1833," and "The origins of Charles Lyell's uniformitarianism." The first volume of Professor Wilson's biography has now appeared: *Charles Lyell: The Years to 1841: The revolution in geology.*

[3] Lyell, *Principles of Geology* (1830), I, pp. 141–143, and the third edition (London, 1834), II, pp. 272–322.

[4] Lyell, *Principles of Geology* 1st edition, I, p. 144.

[5] *Ibid.*, p. 146.

[6] *Ibid.*

[7] *Ibid.*, Chapters 6–8, pp. 92–143.

[8] *Ibid.*, p. 150.

[9] *Ibid.*, p. 152.

[10] "Transactions of the Geological Society of London, volume 1, second series, 1824," see

513 and 538–539. On this review as an indication of Lyell's early ideas, see Wilson, "The intellectual background in Lyell's *Principles of Geology*," 436.

[11] Lyell, *Principles of Geology*, I, p. 155.

[12] *Ibid.*

[13] *Ibid.*, pp. 155–156.

[14] Lyell to Sir J.F.W. Herschel, June 1st, 1836, see K.M. Lyell, *The Life, Letters and Journals of Sir Charles Lyell, Bart.* (London, 1881), I, pp. 467–468.

[15] See the letter of Lyell to Babbage, May 1837, *ibid.*, II, pp. 9–10.

[16] Michael Bartholomew, "Lyell and and evolution: an account of Lyell's response to the prospect of an evolutionary ancestry for man."

[17] Charles Lyell, presidential address (1851), xxxix.

[18] *Ibid.*, xxxv–xl. Brongniart's *Tableaux des genres des végétaux fossiles* had also appeared as the article "Végétaux fossiles" in D'Orbigny's *Dictionnaire universel d'histoire naturelle* of 1849.

[19] Lyell, presidential address, xxxvii, see also lii. Lyell was here relying on Owen's 1846 classification of the fishes, discussed below.

[20] *Ibid.*, liii–liv.

[21] *Ibid.*, lvi–lvii. Lyell had examined the footprints discovered by King in the rocks of Pennsylvania, see his *Second Visit to the United States of North America*, II, pp. 229–235.

[22] Lyell, presidential address, lix.

[23] *Ibid.*, lix–lxii. Hitchcock's main account of the tracks, his *Ichnology of New England*, was published in 1858, but at the time Lyell was writing Hitchcock had already published papers describing them and attributing them to birds.

[24] Lyell, presidential address, xliv–xlv.

[25] *Ibid.*, lix–lxii.

[26] See W.E. Logan, "On the occurrence of a track and footprints of an animal in the Potsdam sandstone of Lower Canada." Owen's opinion was appended to this paper; see his "Description of the impressions in the Potsdam sandstone discovered by Mr. Logan in Lower Canada." He also communicated his interpretation to Lyell in a letter which was quoted at the end of the 1851 presidential address, lxxv–lxxvi.

[27] See Owen, "Description of the impressions and footprints of the Protichnites from the Potsdam sandstone of Canada," which followed Logan's second paper on the topic in 1852.

[28] G.A. Mantell, "Description of the *Telerpeton elginense*, a fossil reptile recently discovered in the Old Red Sandstone of Moray. . . ."

[29] See Sir A. Geikie, *Life of Sir Roderick I. Murchison*, II, p. 120.

[30] This note is also reprinted at the beginning of the fifth (1872) edition of *Siluria* cited in Chapter 5.

[31] This was the *Microlestes antiquus* of Plieninger, from the Rhaetic beds of the upper

Trias in Germany (now known as *Thomasia antiqua*). Owen and a number of other workers regarded it as a marsupial.

[32] Richard Owen, "On some fossil reptilian and mammalian remains from the Purbecks."

[33] For a discussion of Lyell's reactions to these discoveries, ses Leonard G. Wilson's introduction to his edition of *Sir Charles Lyell's Scientific Journals on the Species Question*, pp. li–liv.

[34] See *ibid.*, especially p. 414, pp. 422–423, and p. 458.

[35] See Sir Charles Lyell, *Geological Evidences of the Antiquity of Man, with remarks on theories of the origin of species by variation*, p. 505, where it is suggested that the step from the animals to man may have been "cleared at one bound." Darwin told Lyell that this sentence made him "groan"; see the letter of March 6th, 1863, Francis Darwin (ed.), *The Life and Letters of Charles Darwin*, III, p. 12. Lyell's own letters also make it clear just how disturbed he still was on the issue of man; see those to J.D. Hooker, March 9th, 1863, and to Darwin, March 11th; *Life, Letters and Journals of Lyell*, II, pp. 361–364.

[36] Lyell, *Antiquity of Man*, p. 405.

[37] Lyell to Darwin, May 5th, 1869, *Life, Letters and Journals of Lyell*, II, p. 442.

[38] G.A. Mantell, *The Wonders of Geology, or a familiar exposition of geological phenomena*, II, p. 447 and pp. 778–779.

[39] Edward Forbes, "On the manifestation of polarity in the distribution of organic beings in time," see 429. The theory is also outlined in Forbes' 1854 presidential address to the Geological Society, especially lxxvii–lxxxi. For a brief modern description, see Cannon, "The uniformitarian-catastrophist debate," 51–53.

[40] Forbes, "On the manifestation of polarity," p. 430. Although Forbes regarded the small number of new forms appearing in the middle of the fossil record as part of a divine plan, other naturalists connected the phenomenon with A.C. Ramsay's theory of Permian glaciation, on which see Sir Archibald Geikie, *Memoir of Sir Andrew Crombie Ramsay*, pp. 228–229 and 368–369.

[41] Constant Prévost, "De la chronologie des terrains et du synchronisme des formations," see 1070.

[42] *Ibid.*, 1070–1071.

[43] Constant Prévost, "Quelques propositions relatives à l'état originaire et actual de la masse terrestre, à la formation du sol, aux causes qui ont modifié le relief de sa surface, aux êtres qui l'ont successivement habité," see p. 466–467. On Prévost's position, see Hooykaas, "Geological uniformitarianism and evolution," 12–19.

[44] Prévost, "Quelques propositions," p. 468.

[45] See for instance Alcide Dessalines D'Orbigny, *Paléontologie francaise. Description zoologique et géologique de tous les animaux mollusques et rayonnés fossiles de France. Terrains crétacé*, I, pp. 429–430, and *Terrains oolithiques ou jurassiques*, I, p. 623. On the reception of D'Orbigny's over-rigid stratigraphy see William J. Arkell, *The Jurassic System in Great Britain*, pp. 1–37.

[46] D'Orbigny, *Terrains crétacé*, IV (1847), pp. 371–372.

[47] D'Orbigny, "Recherches zoologiques sur l'instant d'apparition, dans les ages du monde, des ordres d'animaux, comparé au degré de perfection de l'ensemble de leurs organes."

[48] *Ibid.*, 233–234.

[49] *Ibid.*, 234–235.

[50] D'Orbigny, "Recherches zoologiques sur le marche successive de l'animalisation à la surface du globe, depuis les temps zoologiques les plus anciens, jusqu'a l'époque actuelle."

[51] Hugh Miller, *The Old Red Sandstone, or new walks in an old field*, pp. 241–243. This is actually earlier than Agassiz's discussion of recapitulation. The two men had met in 1840, when Agassiz examined Miller's specimens; see Agassiz, "Report of the fossil fish of the Devonian system or Old Red Sandstone." On Miller's life, and his ultimate breakdown and suicide, see Peter Bayne, *The Life and Letters of Hugh Miller*.

[52] Miller, *Old Red Sandstone*, p. 62.

[53] *Ibid.*, pp. 63–64.

[54] *Ibid.*, p. 225.

[55] *Ibid.*, pp. 158–159 and 267–269. For the retraction see, for instance, *The Old Red Sandstone* (1858), pp. 154–155. The discovery of large fossil fish in the Old Red Sandstone was made by the noted amateur geologist Robert Dick, the "baker of Thurso."

[56] *Old Red Sandstone* (1841), 44–45.

[57] Hugh Miller, *Footprints of the Creator, or the Asterolepis of Stromness*, pp. 13–14.

[58] *Ibid.*, Chapter 8, pp. 123–135.

[59] *Ibid.*, p. 179.

[60] *Ibid.*, p. 14.

[61] *Ibid.*, p. 16.

[62] Chambers, *Explanations*, p. 32.

[63] See the letter to Agassiz cited above in Chapter two, note 69.

[64] [Adam Sedgwick], "Vestiges of the Natural History of Creation," see 33, 38, and 54–55.

[65] *Ibid.*, 40–41 and 56.

[66] *Ibid.*, 58

[67] *Ibid.*, 60.

[68] Adam Sedgwick, *A Discourse on the Studies of the University of Cambridge*, 5th edition (Cambridge, 1851), preface, p. cliv and p. ccxvi. The quotation from Sedgwick's 1831 address is on pp. xliv–xlv.

[69] *Ibid.*, p. lvi.

[70] *Ibid.*, p. cvii.

[71] Hugh Falconer, quoted by Sedgwick, *ibid.*, p. cxxvi. An extensive series of Falconer's early descriptions of the Sewalik remains appeared in volume XIX of the *Asiatic Researches. Transactions of the Asiatic Society of Bengal* (1836).

[72] For a survey of the reviews, see Millhouser, *Just before Darwin*, Chapter 5.

[73] Anon., "Geology versus development," see 362 and 368. This review praises Miller's paleontology, but criticizes his attempts to connect the fossil record with revelation.

[74] David Thomas Ansted, _The Ancient World; or, picturesque sketches of creation_, p. 56 and p. 391. G.F. Richardson, _An Introduction to Geology and its associate sciences, mineralogy, fossil botany and palaeontology_, pp. 306–307.

[75] John Phillips, _Life on the Earth, its origin and succession_, p. 191.

[76] See T.H. Huxley, "On the reception of the 'Origin of Species,'" p. 189.

[77] [T.H. Huxley], "Vestiges of the Natural History of Creation," see 336–338.

[78] The history of our knowledge of the dinosaurs has been described by William E. Swinton, _The Dinosaurs_, see especially pp. 21–34. See also Edwin H. Colbert, _Men and Dinosaurs. The search in field and laboratory_; and Justin B. Delair and William A. S. Sargeant, "The earliest discoveries of dinosaurs." For the original definition, see Richard Owen, "Report on British fossil reptiles, part 2," p. 103.

[79] See Owen, "Report on British fossil reptiles," pp. 154–155.

[80] _Ibid._, p. 202. The "perrenibranchiate reptiles" are those amphibians that retain their gills in the adult stage, and are therefore closest to the fish.

[81] _Ibid._, p. 201.

[82] See Richard Owen, _Lectures on the Comparative Anatomy and Physiology of the Vertebrate Animals, part 1, Fishes_, pp. 47–51.

[83] _Ibid._, p. 147. These remarks were quoted by Sedgwick in his _Discourse_.

[84] _Ibid._ See Robert E. Grant, _Outlines of Comparative Anatomy_, p. 56.

[85] Darwin's recollection of Grant is in his "Autobiography," _Life and Letters of Darwin_, I, p. 38. Huxley gave a brief assessment of Grant in his "On the reception of the 'Origin,'" p. 188.

[86] See the Rev. R. Owen, _The Life of Richard Owen_, I, pp. 249–255.

5

Continuous Progression
and the Concept of Divergence

The fossil record had exhibited enormous discontinuities in the 1840s, but new discoveries were being made very rapidly. To some extent these helped to fill a number of the larger gaps that had at first been observed, especially in the case of the reptiles. Perhaps inevitably, those geologists who were most thoroughly committed against transmutation refused to admit that the situation had changed; the aged Sedgwick raised the same arguments against Darwin as he had against Chambers, and there is no doubt that Miller would have done the same had he not broken down and ended his own life in 1856.[1] But those of a more flexible mind began to accept that the new evidence was gradually making continuous development seem more plausible. Darwin cited the case of F. J. Pictet, professor of zoology at Geneva, who had originally argued for discontinuous progression but was having second thoughts by the end of the 1850s.[2] The change of attitude was not brought about solely by the new evidence, although we should not underestimate the very real changes that took place at this level. The belief that the history of life must have been governed by some coherent plan linking all of the known species into a coherent pattern was growing in popularity. But this did not mean transmutation: Agassiz himself had given an early lead in this direction by arguing that although there are no physical connections between species, they can be linked together intellectually to reveal the logical plan of organic development. More and more workers took up this approach as the larger gaps between the fossils began to be reduced. Many followed Agassiz in refusing to accept any kind of transmutation theory, but already there were a few—Owen is the best example—who were prepared to go all the way toward accepting the predesigned evolutionary process suggested by Chambers.

At the same time, however, the increasing complexity of the fossil record began to throw doubts on the more or less linear process of development that Chambers had proposed. In fact, both Agassiz and Chambers in their different ways had imagined that the gradual unfolding of the plan of life led toward man, and this idea had obvious

attractions to those who believed that man was the high point of
creation. Thus Agassiz's American colleague, James Dwight Dana of
Yale, connected the general advance of life with his own theory of
increasing "cephalization"—the process by which the parts of the body
were successively transferred to the services of the brain, which reached
its logical conclusion in man.[3] But from the beginning even Agassiz
had known that the ascent toward man was nothing like a complete
representation of the history of vertebrate life. It was obvious that the
further development of a lower class could have no relevance to the
general progression once the next highest class had appeared. In the
case of the fish, for instance, there were major changes such as the great
expansion of the modern bony fishes occurring in periods long after
the reptiles had taken over the leading position among the vertebrates.
The gradual accumulation of evidence confirmed that the development
of every class had to be treated as an independent phenomenon, largely
unrelated to the advance toward man. In particular it was eventually
realized that the birds, far from being a stage in the ascent from the
reptiles to the mammals, were actually a separate line running parallel
to the mammals and with an origin and history quite unconnected with
the "highest" class.

The whole concept of a linear hierarchy among the classes began
to break down at this point. The process was encouraged by another
factor, again recognized by Agassiz in the 1840s, namely, the
divergence that has taken place within every class.[4] As the evidence
began to yield more and more significant patterns between the fossil
species, it became obvious that the development of each class was not a
linear one. On the contrary, there were many parallel sequences within
the class corresponding to the orders and smaller taxonomic divisions.
Many of these changed in ways that could not be counted as a
progression and some suffered a positive decline in their level of
organization. In the old linear viewpoint the continuity of progression
had been measured by whether or not the higher orders of the class
appeared later than the lower ones. This was still relevant to some
extent—the fact that the amphibians now seemed to have appeared
before any of the orders of true reptiles showed that there had been
some degree of progression within the class. But it was now evident
that in the later history of a class all of the orders developed side by
side and that some of the higher orders might become extinct before
the lower ones, as in the case of the dinosaurs. The transmutationist
would have to assume that at least one line led from a high order
toward the structure of the next highest class, but this element of

progression in the old sense would be isolated among a host of unrelated changes.

The new evidence also began to suggest that the real pattern of development followed by each class involved the divergence of a small number of original forms into many lines leading toward different adaptive modifications. The first members of the class would not only be related to the previous class but would also have a very generalized structure. The group would then divide itself into a number of lines, each becoming more specialized in its own way. Darwin's theory had a natural explanation for this phenomenon of divergence (today we call it adaptive radiation), but the effect itself was already being studied by paleontologists before the *Origin of Species* appeared. Such developments naturally tended to cast doubts on the claim that there was a key line of progress leading toward man. For Agassiz, the divergence within each class would have to be regarded as the working out of variations on each step in the advance up to the human form. But the more divergence became obvious, the less reasonable it seemed to single out the line connecting the classes as the main purpose of the whole process. The anthropocentrism of Agassiz's viewpoint faded, to be replaced by a new version of the utilitarian argument from design. Divergence was the preordained unfolding of the various possibilities to which the class could be adapted. Paley's original claim that the Creator's benevolence could be seen in the perfect adaptation of each form to its environment gave way to the belief that design could be traced out in the process by which the class gradually assumed its more specialized adaptive structures.

Progression and Divergence in British Paleontology

In 1854 Sir Roderick Impey Murchison published a new survey of the Paleozoic rocks under the title *Siluria*. Since he had written the *Silurian System* in 1839 he had been extending his knowledge of these early rocks in different parts of Europe. His efforts to establish that there was "but one natural history group of life in Cambria and Siluria" led to increasing friction with Sedgwick and at last to an open breach of relations in 1852.[5] After some debate Sedgwick was able to maintain the distinct nature of the Cambrian, although eventually the disputed intermediate strata were made into a new system, the Ordovician. Murchison's energies were not, however, expended solely on these matters. His *Siluria* offers an interpretation of the history of life quite different from that of his earlier work, showing that he had now adopted the kind of progressionism originally proposed by Agassiz. He still accepted a directional theory of the earth and admitted that this would influence the order in which living forms were introduced. Thus he argued for the greater intensity of past geological activity on the basis of the cooling-earth theory and explained the greater numbers of terrestrial animals in the later periods as the result of an increase in the amount of dry land. But he was now convinced that the progression of life could not be explained solely in these terms—there must have been a real progressive trend in creation, ensuring that on the whole new species were superior to the ones they replaced. We shall see that in fact Murchison was keenly interested in the overall advance through the classes, but in the first edition of *Siluria* he chose to develop the theme mainly in connection with the fishes.

He began by arguing that the absence of vertebrate fossils in the early Silurian proved (whatever Lyell might say) that this was an age in which invertebrates were the highest forms of life. The cephalopods, he believed, were the predators of these early seas, giving way to the fishes only at the end of the Silurian. The subsequent development of the fishes then served as the principal line of evidence for a progressive trend in creation.

> Just as the introduction of cartilaginous fishes...is barely traceable at the close of the long Silurian era, so becoming soon afterwards more abundant, they are associated in all younger formations with true osseous fishes, whose remains are found intermixed with the other exuviae of the sea. Putting aside, therefore, theory, and judging solely from positive observations, we may fairly infer, first, that during very long epochs the seas were unoccupied by any kind of fishes; secondly, that the earliest discoverable creatures of this class had an internal framework almost incapable of fossilization, and so left in the strata their teeth and dermal skeletons only; and thirdly, that in the succeeding period, the oldest fishes having bony vertebrae make their scanty appearance, but become numerous in the overlying deposits. Are not these absolute data of the geologist clear signs of a progress in creation?[6]

The development of the bony fishes would have been difficult to explain in terms of changing conditions, nor did Murchison make any attempt in this direction. The progression was presented as a distinct trend forming part of the basic plan of creation. But it is clear that he was prepared to see this trend affecting lines of development other than the main one leading toward man. As Agassiz's researches had shown, the real expansion of the bony fishes occurred quite late in the fossil record, long after the reptiles had taken over the lead in the main advance through the classes. The history of life was a multilinear process, and the progressionist had to be prepared to seek his evidence even within changes that were confined to a single class.

In order to establish a progression within the fishes, Murchison had in effect accepted Chambers' interpretation of the low status of the first members of the class. He would not agree, however, that there was a real connection with the previously existing invertebrates. He argued that

> . . . the appearance of the first recognizable fossil fishes is as decisive a proof of a new and distinct creation, as that of the placing of man upon the terrestrial surface, at the end of the long series of animals which characterize the younger geological periods.[7]

There were still enough gaps, in other words, to make transmutation out of the question, but the overall continuity of development suggested a coherent plan of creation. In fact Murchison never shared the violent distaste for *Vestiges* expressed by Sedgwick and Miller.[8] His opposition to transmutation was more conventional and pragmatic, not deriving from any undue distress over the status of man. Miller, for instance, would never have compared the introduction of the fishes to that of man in the tone of the passage quoted above. Murchison thus saw no reason to emphasize the gaps in the fossil record any more than was absolutely necessary to dispose of transmutation. To make the progression completely discontinuous by claiming that the first fishes were the highest would simply destroy the general pattern which gave the chief clue as to the basic nature of God's progressive plan of creation.

The progression within the fishes was the main theme developed in the first edition of *Siluria*, although we know from Murchison's correspondence that he was convinced of the even greater significance of the overall advance through the classes. His reluctance to discuss the wider topic in print may have been at least in part a result of Mantell's discovery of *Telerpeton elginense*, the reptile supposedly from the Old Red Sandstone. He was much annoyed at the support this appeared to give to Lyell's complete anti-progressionism, by making it impossible to establish a real progression from the fishes to the reptiles. All he could do in *Siluria* was to mention *Telerpeton* briefly, trying to fit it into the progressionist picture by claiming that it was more primitive than the later reptiles of the Carboniferous (a somewhat dubious point, as he later admitted).[9] Only in the later editions of the book did Murchison introduce a full discussion of progression, especially after T.H. Huxley had undermined the significance of Mantell's discovery by identifying *Hyperadapteon*, another reptile from the same strata, with a form found elsewhere in the New Red Sandstone. A prefatory note was added to the 1867 edition commenting on the importance of this piece of paleontological detection and on the unusual circumstances that had allowed strata of the New Red Sandstone to lie in conformity with those of the Old.[10] Now Murchison was free to emphasize the overall progression through the classes, which he did by sketching in the history of the fishes, reptiles, and mammals, paying particular attention to the mammals of the Secondary era, which were presented as mainly low forms well in agreement with the progressionist thesis.[11] He concluded

> Let the reader dwell on these remarkable facts which the close labours of the geologists have elicited in this century. Let him view them in the clear and broad order indicated by Nature, advancing from an Invertebrate to a Vertebrate era, and next mark a regular rise thenceforward in the numbers and organization of animals by the addition, in successive epochs, first of Reptiles and then of Mammals. Let him execute a patient survey from the lower deposits upwards, and he will find everywhere a succession of creatures rising from lower to higher organizations,—a doctrine promulgated by the illustrious Cuvier, but from infinitely less perfect data than we now possess.[12]

The belief that Cuvier had founded the theory of progression was common at this time, although in fact the French naturalist had confined himself to establishing the late appearance of the mammals without commenting on the reason for this. Murchison's progressionism was of the new kind, based firmly on the conviction that the fossil record revealed a fundamental plan of organic development aimed at least in part at the ascent toward man. Although he did not emphasize the primacy of the human form in the manner of Agassiz, he did write of the first fish as "the prototype of that vertebrate succession which terminated in man."[13] Yet despite all his enthusiasm for this most important aspect of the progression, Murchison still accepted the idea that the advance was not purely linear. He continued to speak of a development within the fishes leading not toward the next class but toward a particular improvement of the fishes' own structure. If there was a progressive trend in creation, it had to be seen as a multilinear system in which limited advances could be produced in the side branches quite separate from the main line leading toward man.

The implications of the multilinear view of living history were ultimately brought out more clearly by Richard Owen. But Owen, like many others, passed through an early phase in which his growing concern for continuity led him to accept an almost linear approach. In the early 1840s, as we have seen, he had been impressed by the immense discontinuity of the fossil record and convinced that there was no hope of establishing a theory of gradual progression. But his studies in comparative anatomy were inevitably pushing him toward the construction of such a view, and by the end of the decade he seems to have felt that the fossil gaps had been filled in enough to make the generalization plausible. It was his theory of the vertebrate archetype that provided the intellectual driving force—a transcendental vision of the unity of the type in which each species was seen as a more or less complex adaptive manifestation of an ideal form. This theory, published in his *On the Archetype and Homologies of the Vertebrate Skeleton* of 1848, was not strictly speaking a "man-centered" view of the vertebrates in quite the same sense as that supported by Agassiz. [14] The archetype was the simplest and most general vertebrate form, with the significance of man arising only out of the fact that he seems to be the most complex manifestation of the type. But such a view still allowed the construction of a hierarchical arrangement of forms within the class, ideally suited for the basis of a progressionist theory of the ascent of life toward man.

Such a theory was finally suggested in the conclusion of Owen's *On the Nature of Limbs* of 1849. The chief aim of this work was to argue that the utilitarian version of the argument from design associated with Paley's name did not, in fact, provide the most powerful evidence of God's influence on nature. More significant was the fact that the limbs of forms adapted to a host of different circumstances are all built on the same basic plan. This cannot be a coincidence—it is a unity which reveals that the whole type was created by a rational Mind. Following the logic of Agassiz's transcendentalism, Owen then went on to argue that the hierarchical pattern represented by the increasing complexity of the individual modifications must also have formed the historical plan of creation. But unlike Agassiz, he was prepared to follow Chambers in assuming that the plan may have been allowed to unfold itself gradually through the creative forces that we call "laws."

> The archetypical idea was manifested in the flesh, under divers such modifications, upon this planet, long prior to the existence of those

animal species that actually exemplify it. To what natural laws or secondary causes the orderly succession and progression of such organic phenomena may have been committed we are as yet ignorant. But if, without derogation of the Divine power, we may conceive the existence of such ministers, and personify them by the term 'Nature,' we learn from the past history of our globe that she has advanced with slow and stately steps, guided by the archetipical light, amidst the wreck of worlds, from the first embodiment of the Vertebrate idea under its Ichthyic vestment, until it became arrayed in the glorious garb of the human form.[15]

Naturally Owen began to think of revising his opinions of the fossil record to fit in with this concept of living development. An opportunity arose when he was asked to comment on Lyell's 1851 presidential address for the *Quarterly Review*. His critique was published anonymously, but Lyell himself knew that Owen was the author[16] and this knowledge was widespread enough for Chambers to attribute the review to him in the 1853 edition of *Vestiges*.[17] Owen devoted a good deal of space to ridiculing Lyell's claim that the higher animals could have lived in the early geological periods without leaving fossil remains.[18] But Lyell had also supported the view that the first members of each class were often of a high level of organization, an argument that Owen now wanted to attack in the name of continuity. He briefly dismissed the anti-progressionist account of the development of plants,[19] but inevitably devoted most of his space to the animals. Beginning with the fishes, he now reversed in part the opinions he had expressed in 1846. He admitted that the class could not be represented as a simple hierarchy, since many forms combine a high degree of development in one character with a primitive level in others.[20] But driven by his theoretical interest in the skeleton, he now declared that this factor outweighed all others in the assessment of a form's level of development, and hence that a cartilaginous skeleton was an absolutely primitive sign. Since the first fish were cartilaginous, it followed that there has been a progression within the class.

> We have adverted to the remarkable fact that no completely ossefied vertebra of a fish has been discovered in the strata of the Silurian and Devonian period. Those strata are of enormous extent, and have been most extensively investigated. As regards the internal skeleton, these primeval fish were less fully developed that those of the tertiary and existing seas.[21]

Owen was now prepared to give the superficial resemblance of the early forms to the invertebrates at least some significance, and he also laid considerable emphasis on the resemblance of the first fishes to the embryonic form of their modern counterparts through their possession of a heterocercal tail.[22] The class could thus be said to have undergone a general progression from its earliest, most primitive forms. But it may be noted that Owen was already admitting that the progression was multilinear, since he measured the advance not so much in terms of a move toward the structure of the next highest class, but by the later production of the modern bony fishes.

Passing on to the reptiles, Owen again had to reverse his original opinion, as expressed in his 1841 report. This time, however, he had sound evidence from new discoveries on which to base his arguments— indeed it is not impossible that the changing situation here was a leading factor in convincing him that continuous progression was now plausible. Lyell had himself examined some reptilian footprints from the Carboniferous rocks of Pennsylvania, and Owen now argued that these appeared to have been made by amphibians.[23] This fitted in well with the progressionist interpretation of the reptiles' development, but even stronger evidence had now been provided in the form of skeletons unearthed from the coalfields of Germany. The first of these forms had been described by Herman von Meyer under the name *Apateon terrestris* (earlier incomplete specimens had been regarded as fishes rather than reptiles). Three further forms were discovered in the coalfields of Saarbruck and were given the generic name *Archegosaurus* by Goldfuss in 1847. (See Plate VIII.) Goldfuss had regarded these species as related to the crocodiles, but von Meyer argued that they were closer to the Labyrinthodonts of the Trias, which he in turn thought were related to the Sauria or lizards.[24] This would have made a progressionist interpretation impossible, but von Meyer later changed his opinion to suggest that *Archegosaurus* was a reptile arrested in its larval stage of development, resembling the amphibians of today.[25] Owen himself regarded the Labyrinthodonts as structurally intermediate between the modern amphibians and reptiles, and he now accepted von Meyer's new opinion of *Archegosaurus*.[26] He noted a resemblance between these forms and the modern perrenibranchiate amphibian *Proteus* (which keeps its gills when adult), coupled with a highly developed "dermal skeleton" which seemed to indicate a relationship with the first fishes. The reptiles thus appeared in a very primitive form and progressed up to the dinosaurs, which Owen still treated as the highest order within the class.

Concluding with the mammals, Owen first discussed the remains from the Secondary era on which Lyell placed so much emphasis. While admitting that *Amphitherium* now had to be regarded as an insectivore rather than a more primitive marsupial, he still claimed that this left the class with only a limited level of development until the Tertiary. Lyell had also pointed out a brief passage in Owen's earlier *British Fossil Mammals* where it had been suggested that mammalian carnivores must have existed to control the population of the more primitive Secondary mammals, but Owen now simply repudiated this by pointing out that the carnivores might just as well have been reptiles.[27] In his discussion of the mammalian development during the Tertiary, Lyell had argued that there was no sign of a gradual advance toward man. Owen set out to counter this in two ways: first by challenging Lyell's basic assertion, and second by redefining progression so that it was no longer a purely anthropocentric doctrine. On the first line of attack he pointed out that there did seem to be an increased proportion of the higher mammalian orders in the later periods of the Tertiary, particularly of the ruminants and the quadrumana (the apes).[28] The redefinition of progress arose out of the comparatively well-known fact that many of the ancient mammals seemed to combine characters now separated among distinct modern families. The Eocene *Palaeotherium,* for instance, had affinities with both the modern horse and the tapir.[29] In effect it was possible to trace out a process of specialization within the class, as structures such as the horns, feet, and teeth gradually departed from the most general or archetypical mammalian structure. The three toes of *Palaeotherium* thus gave way to the more specialized single hoof of the modern horse. This process could be equated with a progression, indirectly because man himself is the least archetypical mammal, and directly since the specialization conferred distinct adaptive advantages on the animals concerned.

In his concern to establish a mammalian progression, Owen had drastically altered the criteria to be used in assessing the level of organization and had abandoned his 1849 speculation about a steady ascent toward man. The development of the class was now seen as a multilinear process with many lines diverging away from the early, more generalized types toward different adaptive modifications. Such a view could only very indirectly be related to old idea of a progressive trend toward man, and in effect the whole hierarchical view of fossil record was now being threatened. Owen himself was to become increasingly conscious of this in the course of the decade, while other workers also began to sense the importance of a new approach along these lines. A leading figure here was William Benjamin Carpenter, one of Britain's most respected physiologists. Carpenter played a key role in the introduction of K.E. von Baer's new embryology into the country, expounding it at some length in the third edition of his *Principles of Physiology* of 1851. He saw the possibility of drawing an analogy between von Baer's concept of development from homogeneity to heterogeneity of structure and the trends that could be seen in the fossil record, in effect, founding a new kind of recapitulation theory not based on the old law of parallelism that von Baer discredited. Carpenter opposed *Vestiges* (although he later accepted evolution in the post-Darwinian period) on the grounds that there were too many cases in which the first known members of a group were not the lowest.[30] He argued that in all cases, however, the earliest members of a class were the most general in structure, and that the subsequent development consisted of the appearance of many forms each emphasizing certain aspects of the original structures. His own paleontological interests lay with the invertebrates, from which he cited the example of the echinoderms, appearing first in the form of the Cystidea, by no means the lowest group but one which encompassed "a most extraordinary combination of the characters of the remaining groups."[31] From the vertebrates he gave the example of the labyrinthodonts, which appeared to combine features later developed separately in the modern amphibians and crocodiles. Although Carpenter's idea that such developments paralleled von Baer's conception of embryological growth did not excite much interest among other naturalists, the illustrations he gave were often those that Owen was also beginning to regard as significant. The same effect was also mentioned in the *Essays on the Spirit of the Inductive Philosophy* (1855) of Baden Powell, Savilian professor of geometry at Oxford. One of Powell's main aims in this forward-looking and controversial work was to argue that evolution was not necessarily incompatible with natural theology. Yet he was very suspicious of the theory of continuous progression, preferring a multilinear view of the history of life in which the most prominent phenomenon was "a combination of characteristics of several species, or even genera, or orders in the same individual in one period, to be developed separately in different species in a succeeding era. . ."[32]

Powell was not himself a paleontologist, so his claim that divergence was the more obvious pattern to be seen in the fossil record suggests that the superficiality of the old progressionism was now becoming widely recognized. The chief exponent of the new approach, however, continued to be Owen, whose attempts to synthesize an overall picture of the history of life continued throughout the decade. In 1860, Owen summarized his thoughts in his *Palaeontology, or a*

systematic study of extinct animals and their geological relations.
Although it contains some comments on Darwin's newly published
theory, this work was evidently not intended as a refutation and in
many superficial ways its conclusions were in agreement with the new
evolutionism. Owen had continued to work on the assumption that the
development of life was an almost continuous process, but even more
than in 1851 he had now realized that the fossil record cannot be
understood in terms of a progression toward man. By far the most
obvious pattern was that of divergence and specialization, and he
seems to have appreciated that there was no longer any point in trying
to equate these processes with the old concept of progression.

Even in his attack on Lyell, Owen had admitted that it was
difficult to classify the earliest known fish as "low" in any absolute
sense. But he had stressed that the possession of a cartilaginous
skeleton was a primitive characteristic that would outweigh the
apparently high status of a number of their other structures. He now
accepted a more equal balance between the various criteria, and was
thus forced to admit the impossibility of treating the historical
development of the class as a simple progression toward a higher form.
He pointed out that the ancient fish that most closely resembled the
reptiles in the structure of their brain and generative system
nevertheless possessed "embryonic" features such as the heterocercal
tail, indicating that they were "arrested in development." Futhermore,
the most reptilian fish were not those which represented the most
extreme of the purely fish-like modifications of the vertebrate type.[33] In
other words, many orders of fish have undergone extensive developments in ways which do not correspond to an advance toward the next
highest class; they have acquired specialized modifications which are
appropriate only to their own class. Owen concluded that "A
retrospect of the genetic history of fishes imparts an idea rather of
mutation than of development . . ." and argued that the present time
must be treated as a "period of mutation of the piscine character,
depending on the progressive assumption of a more special piscine
type, and a progressive departure from a more general vertebrate
type."[34] In effect, he had realized that the only kind of development
that could have been described as progressive in the original sense of
the word would be an advance from the simplest kind of fish up to the
most reptilian. The enormous developments which had taken place
after the reptiles had appeared were clearly irrelevant to the general
advance through the hierarchy of the classes, as indeed would have
been many of the earlier trends. He thus abandoned progressionism

and, as Rudwick has pointed out, began to emphasize that the fossil record exhibits a series of trends in which the most archetypical forms are gradually modified toward an increased degree of adaptive specialization.[35] Adaptation might thus still prove design, but the fossil record showed that Paley's original argument would have to be modified to accommodate the fact that the degree of adaptation became progressively more effective as time went on.

Within the early history of the reptiles themselves, Owen still felt that it was possible to recognize something of a progression in the sequence with which the orders had appeared. He simply ignored Mantell's *Telerpeton elginense*, declaring that the oldest members of the class were the Ganocephala, armored amphibians from the Carboniferous including *Archegosaurus*.[66] The later appearance of higher orders such as the dinosaurs represented a progression, although the class has now declined both in numbers and in its level of organization. Owen believed that the eventual extinction of some of the higher orders was a direct result of changing physical conditions; the progress of the class had been checked because it had become "unequal to the exigencies and life-capacities of the present state of the planet."[37] He made it clear, however, that a simple picture of progress and decline was not an adequate representation of the history of the class. He used a diagram to show that the various reptilian orders could be treated as a series of parallel lines of development. (See Plate IX.) The fact that some of the more primitive orders started out earlier than the others constituted a progression, but this was a comparatively minor phenomenon when compared to the development of the whole range of orders. More significant was the tendency of the earlier forms to be more generalized in structure; the ancient Ganocephala and Labyrinthodontia in particular were of a "more generalized vertebrate structure" than the modern reptiles. Related to this was the fact that some of the earlier reptiles appeared to combine characters now separated out among distinct modern orders. Already in 1855, Owen had noted this tendency in the skull of the South African *Dicynodon*, and now he argued that both the ancient Cryptodontia and Dicynodontia combined structures manifested separately by the modern crocodiles, turtles, and lizards.[38]

The mammals presented an essentially similar picture. Rather more mammalian remains had now been identified from the Secondary era, but Owen still regarded them as limited to the more primitive marsupial and insectivorous orders. The real development of the class could still be seen to take place in the Tertiary, when most of the modern orders appeared and began to develop side by side.[39] Owen no longer seems to have been interested, however, in describing the expansion of the mammals as a progression. He did not draw much attention to the later appearance of the apes and of man, preferring to take the general process of divergence and specialization at its face value as the most significant trend. He argued that the level of diversity of the class has increased in the course of the Tertiary, since in many cases the orders were less distinct when they first appeared, their forms more generalized in structure. Thus the Ruminants, regarded by Cuvier as one of the most well-defined modern orders, could be traced back via the structure of the teeth to the Eocene *Dichodon* and *Anopletherium*, which did not have the specialized ruminating stomach.[40] Other Eocene forms such as *Palaeotherium* exhibited structures which could be related to a number of modern families. Only later, in Miocene and Pliocene times, did forms more closely resembling the distinct modern families begin to appear. At this point Owen introduced a brief mention of Darwin's ideas and stressed that he himself had

> . . . never omitted a proper opportunity for impressing the results of observations showing the "more generalized structures" of extinct, as compared with the "more specialized forms" of recent animals.[41]

Although Owen could not accept Darwin's purely naturalistic explanation of this trend, it is clear that his overall view of the development of life was now a highly sophisticated one, far closer to the modern approach than the old linear progressionism. He still accepted a general increase in the level of organization, but this was no longer related to the idea of an unambiguous ascent toward man. Far more general was the trend toward increasing diversity of structure brought about by the specialization of the orders and families within each class.

Owen's rejection of natural selection and his refusal to admit Huxley's point that man is closely related to the apes have often led to his being dismissed as a straightforward opponent of evolution. But we have seen that at least from 1849 he was quite prepared to accept the continuous development of life brought about by "secondary causes."[42] In the *Palaeontology* he again talked of a "continuously operative secondary creational power,"[43] but his later theory of "derivation" made it quite clear that such powers were to be seen as direct manifestations of the Creator's will, and that they gave evidence of design through their gradual unfolding of the adaptive possibilities inherent in the basic structure of each class.[44] Obviously from such a background he would reject natural selection, but the patterns he observed in the fossil record were very similar to those predicted by Darwin and in fact the *Origin of Species* refers to Owen on this point.[45] Although his review of the *Origin* was confused and vindictive, it is evident that Owen was never prepared to abandon evolution altogether. Even when he criticized Huxley's postulated derivation of the horse, tapir, and rhinoceros from *Palaeotherium* on the grounds of anatomical dissimilarities, he admitted that different possible connections might make the "derivative hypothesis" more plausible.[46] In general his opposition to natural selection did not and could not rest upon his paleontological work, since his disagreement with Darwin was really over the explanation of the same basic patterns. His refusal to accept man's close relationship to the apes—which led to such a damaging debate with Huxley[47]—can perhaps be ascribed to Owen's inability to recognize criticism of what was originally a quite genuine mistake. Certainly he never gave any sign that he suffered from theological qualms over the connection of man with the animals. Owen was always a sensitive and self-centered personality, but never a blind opponent of evolutionism.

Progression and
Divergence in Europe

Although Adolphe Brongniart continued to support progressionism in his *Tableaux des genres des végétaux fossiles* of 1849, French writers in general did not take up the new approach advocated by Agassiz. Geoffroy's transcendentalism came under increasing attack from naturalists convinced of the diversity of natural forms,[48] and the possibility of distinguishing a general plan in the history of life seems to have been largely ignored. The Paris Académie des Sciences proposed a prize in 1850 for the best study of the laws of development of the organic world, but it was not awarded until the end of the decade and then to a German: H.G. Bronn. It was, in fact, the Germans who became increasingly concerned with such issues and, as Owsei Temkin has described, there were numerous attempts to explore even the possibility that the structure of nature could have unfolded without the necessity for miraculous intervention.[49] The paleobotanist Franz Unger, for instance, at first suggested a series of linked spontaneous generations to account for the pattern of development, but finally switched to a transmutation theory. As Lovejoy has noted, even the philosopher Schopenhauer promoted the idea of evolution through a series of mutations in 1851.[50] Almost always such systems were based on the conviction that the history of life was governed by some kind of coherent plan—hence perhaps the popularity of *Vestiges* in Germany. To this extent the spirit of *Naturphilosophie* was still influential, despite the formal rejection of the mystical approach by most German scientists at this time. As Temkin's discussion shows, however, most of the new systems of development avoided the old tendency to treat the whole of nature as centered on or striving toward man. Many workers now recognized that the ascent of life was a multilinear process, not an evolution through a single hierarchy with man at its head. Perhaps the clearest attempt by a German naturalist to base such a view of nature firmly on the fossil evidence itself occurs in the writings of H.G. Bronn.

Heinrich Georg Bronn had been appointed professor of natural science at Heidelberg in 1833. Soon after this he began the publication of his *Lethea Geognostica*, which established itself as a standard paleontological reference work. Continuing his efforts in this field he produced in the late 1840s an *Index Palaeontologicus* detailing all of the known fossil forms. He then went on to analyze the laws governing the successive appearance of new living forms, publishing his conclusions in his *Untersuchungen über die Entwickelungsgesetze der organischen Welt* in 1858. This work was also submitted to the Paris Academy in response to the prize offer of 1850, and was thus printed in French both as a monograph and as a supplement to the *Comptes rendus*.[51] Russell has described Bronn's general approach to nature as that of a transcendentalist,[52] but his was certainly not a man-centered system. On the contrary, he actively sought the causes which produce organic diversity and was quite prepared to accept adaptation to changing conditions as a leading factor giving rise to the divergence occurring in the fossil record. His work thus contributed to the general trend by which man was removed from his position at the head of the vertebrate hierarchy and progression defined in a more sophisticated way.

Bronn could not support transmutation, since he held that the gaps between the known fossil forms were too large to be compatible with complete continuity of development.[53] Each species must have been created separately, but it was evident that the succession of forms could be linked together to reveal coherent lines of development. Bronn was sufficiently interested in the possibility of transmutation to supervise a German translation of the *Origin of Species*, although he was not actually converted to Darwinism. His approach was to study the basic laws (trends would perhaps be a better word) underlying the observed sequence of fossils, interpreting these as elements of the predetermined plan of living development. He proposed two such laws; the first dictating the necessity of all forms to be adapted to their environment, the second a gradual progressive advance in the level of organization. The first law itself led Bronn to a directional theory of the history of life, since he was committed to directionalism in geology and held it to be obvious that the organic and physical worlds must have a parallel development. A decline in the carbon dioxide content of the atmosphere may have exerted some effect during the early geological periods.[54] The gradual cooling of the earth has also been of some significance, since the creatures from the lower part of the record are all suited to a hot climate. But Bronn explicitly denied that this factor could explain the low level of organization of these early forms—he was not prepared to follow what we have called the old form of progression favored by the British catastrophists.[55] Perhaps the most important directional change followed the gradual expansion of the land surface through geological time. To this corresponded a law of "terripetal" development, according to which living forms were gradually created with a greater capacity for a terrestrial mode of life.[56]

The law of terripetal development may have harmonized with the observed progression of vertebrate life, but Bronn again made it clear that such a factor could not serve as a complete explanation of the advance in organization. There was, in addition, "a progressive development of successive populations following a law independent of the external conditions and inherent to the creative force itself."[57] Left to itself, this law would have produced a "simple and uniform" ascent of an organic hierarchy of complexity, but the law of adaptation had constantly interfered with this.[58] Bronn realized that this interference has divided the history of life into a multitude of different lines of development, each advancing in its own way. Thus the geological distribution of every class could be represented as a series of parallel lines corresponding to the separate orders.[59] In fact the overall history of life could be described as a process resembling a tree in which new branches are always splitting off in the course of the ascent, as illustrated by the diagram reproduced here. (See Plate X.) There was still meant to be a general progression underlying all of the many developments: Bronn explained that the vertical axis of his figure represented both time and the level of organization, at least as far as the point of origin of each line was concerned. (That is, the higher the point of origin the higher the level of organization, but continued extension of a line in an upward direction indicates only duration in time, not that an earlier line may come to equal a later one in its level of complexity.) Once created, a particular line of development shows at best only a limited kind of progression, and it is thus possible for a class as a whole to undergo a decline. In the reptiles, for instance, some of the earlier forms had a much greater similarity to the mammals than those of today; evidently these forms had died out, leaving only the lines corresponding to lower levels of reptilian organization to continue through to the present.[60] Although Bronn's diagram seems to have a privileged axis running through it that could easily have been treated as a key line running toward man, he made no effort to expound such an interpretation in his text. Progression could no longer be seen as the gradual advance through a linear hierarchy. On the contrary, it was a much more general trend that could affect all of the various lines of development in a manner totally unrelated to the advance toward man.

The view of living development suggested by such workers as Bronn and Owen was revolutionary in many ways. First of all, it drastically reduced the size of the discontinuities in the fossil record, bringing it more into line with what the transmutationists predicted. Even those who shared Bronn's distrust of transmutation itself helped to undermine the arguments against it which had been based on discontinuous progression. Where Sedgwick and Miller had been able to crush Chambers by citing the almost universal conviction that the first members of each class were the highest, Darwin could afford to relax a little on this issue since some major paleontologists were now on his side. Of course the record was full of lesser discontinuities and Darwin still felt the need to explain these away. But the vast irregularities postulated by earlier creationists had now vanished and the real problem that remained was the more technical one of the lack of intermediates between the already known forms. It was easier to dismiss this as a result of the imperfection of the fossil record, and Darwin's efforts in this direction met with some degree of success. At least, paleontology played a much less important role in the debate over the *Origin of Species* than it had in the refutation of *Vestiges*.

At the same time Owen and Bronn had created a totally new vision of the history of life, combining in an almost unrecognizable form elements of the two previous interpretations. Originally, the British catastrophists had followed Paley's utilitarian version of the argument from design and assumed that each fossil species was perfectly adapted to the environment of its time. Attempts to seek meaningful connections between successive forms had been discouraged, apart from the very general claim that directional changes in the earth's physical conditions had produced an overall progression. Now it had become increasingly obvious that there *were* meaningful links within the sequences of fossils. It was possible in some cases to put together a series of related forms which displayed a gradual specialization of a particular structure. Adaptation thus could no longer be a state of perfect harmony with the environment—it was a *process* which could be observed in the gradual (if not quite continuous) acquisition of more specialized structure within a host of parallel lines of development in the fossil record. But this approach had also demolished the linear progressionism first proposed by Agassiz as an alternative to the utilitarian concept of design. For Agassiz the succession of forms could be linked together to make a coherent pattern whose central theme was the ascent toward man. Chambers' "law" of development was really just the mysterious but gradual unfolding of such a plan. Now the extent of living divergence had been recognized, however, such an approach had to be replaced by one in which the "laws" represented much more loosely structured trends. The law of divergence and specialization followed no harmonious pattern; it simply dictated the expansion of all the adaptive possibilities by which the basic structure of the class could be reconciled with the conditions of a particular era. When combined with this, progression also lost its status as the key element in a transcendental, man-centered plan of creation. It too became just a general trend affecting many different kinds of structure, not an advance along a predesigned and more or less linear hierarchy. Nor was the law completely universal—there were instances where degeneration took place against the general background of progression. The lines which now appeared to link the succession of forms could not be combined to give an overall pattern, least of all one aimed mainly at the eventual production of man.

Although the paleontologists had drastically altered both original versions of the argument from design, for the most part they were still prepared to accept that the new kind of law was a product of the divine will. The trends observed in the fossil record represented the Creator's constant efforts to adapt life to changing conditions while at the same time raising the overall level of organization. Despite the superficial similarity between their view of living development and Darwin's theory, the majority of paleontologists were still emotionally unwilling to accept natural selection or any other purely naturalistic explanation of the trends they observed. Having condemned the old evidence for design, they were nevertheless committed to an attempt to reconcile the new discoveries with what could be salvaged from the old theological principles. Paleontology had become a much less useful argument against a transmutation theory such as Darwin's, but a debate over the true driving force of living development was still inevitable.

Whatever the limitations of the way in which workers like Bronn and Owen tried to interpret the history of life, their efforts represented a notable step toward the concept of a law-governed organic world and paved the way, at least indirectly, for Darwin's total rejection of design. There was clearly a growing willingness at this time to look for laws or trends in the fossil record, strong enough to encourage a major reinterpretation of the status of the early fish even when no new evidence was available. Yet the role of new evidence in promoting this trend cannot be underestimated. Would Owen, for instance, have completely reversed his earlier opinions had not the growing fossil support for a reptilian progression helped to back up his theoretical disposition toward continuity? Furthermore, it seems highly probable that it was the increasing complexity of the fossil record that forced paleontologists to abandon Agassiz's vision of a key line of development in favor of the principle of divergence. The old linear progressionism simply broke down under the weight of evidence showing that a host of different forms had branched out in all directions within each class. It has become unfashionable nowadays to speak of scientists making major advances purely on the basis of the raw evidence they collect. Yet here we have unusual circumstances where the rate of new discoveries produced a dramatic increase in the complexity of the known fossil record. In this case it seems not unreasonable to assume that the evidence itself forced the majority of naturalists to re-evaluate the old ideas about design, progression, and the linear approach to taxonomy. Although this was not enough to demolish belief in the divine control of nature, it *was* enough to make naturalists revise and qualify their views on the nature of that control. The new laws of development did not give such immediate evidence of design, and indirectly at least this paved the way for the acceptance of Darwin's completely naturalistic theory.

Notes

[1] See Sedgwick's letter to the *Spectator* (24th March 1860), 285. He also delivered a paper to the Cambridge Philosophical Society which does not appear to have been published;

see J.W. Clark and T.M. Hughes, *The Life and Letters of the Rev. Adam Sedgwick,* II, pp. 361–362.

[2]See Charles Darwin, *On the Origin of Species* (1859), p. 335. Pictet had supported discontinuous progression in his *Traité élémentaire de paléontologie, ou histoire naturelle des animaux fossiles considerés dans leurs rapports zoologiques et géologiques,* I, pp. 70–80. Darwin's claim that he had changed his mind was only partially supported by Pictet's review of the *Origin*: "Sur l'origine de l'espèce par Charles Darwin," translated in D.L. Hull, *Darwin and his Critics. The reception of Darwin's theory of evolution by the scientific community,* pp. 142–154. Within a few years, however, Pictet had become a complete convert to transmutation.

[3]See for instance Dana's review of Agassiz's *Essay on Classification.* On Dana's attitudes in general see William F. Sanford, Jr., "Dana and Darwinism"; and Morgan B. Sherwood, "The Dana-Lewis controversy, 1856–57. Genesis, evolution and geology." It is perhaps worth noting that Agassiz himself turned back to discontinuous progression at the end of his career in a desperate attempt to undermine the now triumphant evolutionism; see his posthumously published "Evolution and permanence of type."

[4]See the diagram noted above in Chapter three, from *Poissons fossiles,* I, facing p. 170, which clearly shows the families within each order branching away from one another.

[5]See Murchison's letter to Sedgwick of 22nd February 1852, *Life and Letters of Sedgwick,* II, pp. 218–219. This letter was written to justify the Geological Survey's decision to color the Cambrian as though it were Silurian on its latest map, the incident which brought the dispute to a head.

[6]Sir Roderick I. Murchison, *Siluria. The history of the oldest known rocks containing organic remains, with a brief sketch of the distribution of gold over the earth,* p. 462.

[7]*Ibid.*, p. 461.

[8]See for instance the letter written to Owen asking him to prepare a critique of *Vestiges,* in *Life of Owen,* I, p. 254. This is much milder in tone than the letter from Sedgwick making the same request.

[9]Murchison, *Siluria,* p. 462.

[10]See the note dated 1867 prefaced to the 5th edition of *Siluria* (1872) and also pp. 266–267. Murchison himself had examined the strata from which *Telerpeton* and *Hyperadapteon* were derived and had been unable to establish any geological reason why it should be separated from the neighboring Old Red Sandstone. Huxley had at first assured him that these reptiles were quite distinct from any known Mesozoic forms and it was only the later discovery of more remains of *Hyperadapteon* in rocks known to belong to the New Red Sandstone which allowed him to reverse this opinion.

[11]*Ibid.*, pp. 481–488.

[12]*Ibid.*, p. 484.

[13]*Ibid.*, p. 481.

[14]For a description of Owen's archetype theory, see for instance E.S. Russell, *Form and Function,* pp. 104–107.

[15]Richard Owen, *On the Nature of Limbs. A discourse delivered on Friday, February 9, at an evening meeting of the Royal Institution of Great Britain,* p. 89.

[16]See Lyell's letter to Owen, *Life of Owen*, I, pp. 372–373.

[17][Chambers], *Vestiges of the Natural History of Creation*, 10th edition (1853), pp. xiii–xiv. Huxley poked fun at Owen in his review of this edition of *Vestiges* by claiming that Chambers' attribution of the 1851 review to him must have been mistaken, since the opinions expressed therein were so contrary to those of the *Lectures on Comparative Anatomy*; see Huxley, "Vestiges of the Natural History of Creation," 338. Huxley knew, of course, that Owen really was the author; see Leonard Huxley, *The Life and Letters of Thomas Henry Huxley*, I, p. 136 and p. 243.

[18][Richard Owen], "Lyell on life and its successive development," see 436–445.

[19]*Ibid.*, 420–421.

[20]*Ibid.*, 425.

[21]*Ibid.*, 426.

[22]*Ibid.*, 426–429. Huxley pointed out in his review of *Vestiges* that these two criteria were mutually self-contradictory.

[23]*Ibid.*, 422–423.

[24]See von Meyer's review of Goldfuss' *Beiträge zur vorweltlichen Fauna des Steinkohlen gebirges*, translated under the title "The reptiles of the coal formation" in the *Quart. J. Geol. Soc. Lond.* (1848).

[25]See von Meyer, "On the *Archegosaurus* of the coal formation."

[26]Owen, "Lyell on life," 423.

[27]*Ibid.*, 442. For Owen's original opinion see his *History of British Fossil Mammals and Birds*, p. xiv.

[28]Owen, "Lyell on life," 446–447.

[29]*Ibid.*, 449.

[30]William B. Carpenter, *Principles of Physiology, general and comparative*, 3rd edition, p. 577. It was by reading this work that Herbert Spencer first became aware of von Baer's principle of development, later to become a key factor in his evolutionary philosophy; see his *Autobiography*, I, p. 201. On Carpenter's life see the introductory memoir by J. Erstlin Carpenter in W.B. Carpenter, *Nature and Man. Essays scientific and philosophical*. For a detailed discussion of the effects of von Baer's concept of development on paleontological thought, see the article by Dov Ospovat, forthcoming in *J. Hist. Biology*, Spring, 1976.

[31]Carpenter, *Principles of Physiology*, pp. 578–579.

[32]Baden Powell, *Essays on the Spirit of the Inductive Philosophy, the unity of worlds and the philosophy of creation*, p. 333.

[33]Richard Owen, *Palaeontology, or a systematic study of extinct animals and their geological relations*, pp. 150–151.

[34]*Ibid.*, p. 150.

[35]See Rudwick, *The Meaning of Fossils*, pp. 207–214.

[36]Owen, *Palaeontology*, pp. 168–183.

[37]*Ibid.*, p. 284.

[38] See *ibid.*, and Richard Owen, "Description of the skull of *Dicynodon* (*D. tigriceps*, Ow.), transmitted from South Africa by A.G. Bain, Esq."

[39] See diagram in *Palaeontology*, p. 407.

[40] *Ibid.*, pp. 332–335 and 368–370.

[41] *Ibid.*, p. 406.

[42] For a survey of Owen's opinions see Roy M. MacLeod, "Evolutionism and Richard Owen."

[43] Owen, *Palaeontology*, p. 406.

[44] On the theory of derivation see Owen, *On the Anatomy of the Vertebrates*, III, pp. 807--809.

[45] See Darwin, *On the Origin of Species*, pp. 329–330.

[46] See [Owen], "On the Origin of Species," 522–523. Owen himself speculated on the connection between *Palaeotherium* and the horse in the *Anatomy of the Vertebrates*, III, pp. 791–793.

[47] For Huxley's description of this debate see his *Evidences as to Man's Place in Nature*, pp. 113–118.

[48] For a more detailed discussion of these developments see Russell, *Form and Function*, pp. 190–212.

[49] Owsei Temkin, "The idea of descent in post-romantic German biology."

[50] A.O. Lovejoy, "Schopenhauer as an evolutionist."

[51] H.G. Bronn, *Untersuchugen über die Entwickelungsgesetze der organischen Welt während der Bildungzeit unserer Erdoberfläsche* (Stuttgart, 1858). I have used the French version which appeared under the title *Essai d'une réponse à la question de prix proposée en 1850 par l'Académie des Sciences . . ., savoir:—Etudier les lois de la distribution des corps organisés fossiles . . .;* also *Comptes rendus (Supplement)* II (1861), 377–918. The concluding summary of the work was translated into English as "On the laws of evolution of the organic world during the formation of the crust of the earth," *Ann. and Mag. of Nat. Hist.* (1859).

[52] See Russell, *Form and Function*, pp. 202–204.

[53] Bronn, *Essai*, monograph p. 139, *Comptes rendus* 515.

[54] *Ibid.*, monograph pp. 169–173, *Comptes rendus* 545–549.

[55] *Ibid.*, monograph p. 176, *Comptes rendus* 552.

[56] *Ibid.*

[57] *Ibid.*, monograph p. 192, *Comptes rendus* 568.

[58] *Ibid.*, monograph p. 143, *Comptes rendus* 519.

[59] *Ibid.*, monograph p. 522, *Comptes rendus* 898.

[60] *Ibid.*, monograph p. 496, *Comptes rendus* 872.

6

Darwinism and Progression

In the mid-nineteenth century the hierarchical arrangement of the vertebrate classes had become the key to the understanding of the history of life. Where the early catastrophists had merely related the successive appearance of higher forms to the earth's physical development, Agassiz and Chambers had in their different ways explored the possibility that the very essence of nature's historical plan was the ascent through the classes toward man. Owen had at first been excited by this same approach, but by 1860 had become a leading figure in the movement to reinterpret the hierarchical viewpoint. The sheer complexity of living development made it impossible to see man as the inevitable goal of creation, impossible even to be sure that all evolutionary lines led toward increasing complexity. Progression had become an irregular trend underlying the general expansion of life, pushing most branches of the development toward higher levels of organization within their own kind of structure. Adaptation by increasing specialization could now be seen as the real basis of the gradual diversification of living forms. Into this already highly modified climate of thought, the *Origin of Species* introduced a whole new approach to the study of living development. Avoiding the issue of progression, Darwin had been led straight into a search for a mechanism to explain the continuing adaptation of species to a changing environment. Unlike the majority of his contemporaries, he sought directly for a naturalistic explanation that would owe nothing to the Creator's immediate activity and benevolence. From the theory of natural selection he was inevitably driven to predict that divergence and specialization would be important trends in the history of life. Owen's work supplied useful support here, along with his contention that the fossil record could be interpreted in terms of largely continuous trends. Like many others, however, Owen refused to accept that the continuing adaptation to new conditions was achieved without the Creator's direction and his later theory of "derivation" was just one of the many attempts that were made to assert or to prove the insufficiency of natural selection. But in a sense time was very much on Darwin's side: once it was established that there were adaptive trends in the fossil record Paley's argument from design was shaken to the

117

core. Continued efforts would be made to modify it to fit the new state of affairs, but in the long run younger naturalists especially preferred Darwin's theory which explained these trends quite naturally. Although Darwin was still very worried about the fossil record—and it is true that at the detailed level there were still many problems and discontinuities—paleontology was never developed into a leading argument against his theory. Some degree of continuity was assured, and the patterns of development now recognized by leading paleontologists were just what the new theory predicted.

There remains the question of progression. In principle, the new theory simply sidestepped the issue by concentrating on adaptation as the sole key to the evolutionary process. This indeed has been the ultimate consequence of the gradual acceptance of Darwinism—T.H. Huxley began the process in the 1860s and his grandson has led the chorus of voices proclaiming the obsolescence of the old progressionism in the twentieth century.[1] But it would be an underestimation of the Victorians' general faith in progress to expect that the issue could be abandoned as easily as this. Just as Owen and Bronn managed to retain a role for progression despite their recognition of the significance of divergence, so many of the Darwinians continued to believe that their own quite different view of the underlying causes might still imply an indirect advance in the history of life. Darwin himself was highly suspicious of progression and of the whole attempt to rank organisms against a scale of complexity. Yet he could not escape the feeling that the naturalist did have a sense by which he recognized degrees of complexity and that an increase as measured by this sense would—albeit indirectly—be a result of the operations of natural selection. Paleontology had certainly shown that man could not be seen as the logical goal of creation, but it was still clear that the human species does stand out beyond all others through its possession of intelligence. Darwin maintained that this was itself developed by natural selection, the highest product of the whole evolutionary process. Man took up a new position at the head of creation, not as the transcendental key to the whole process of development, but as the form which has been pushed further than all the others by the general progressive trend, into a whole new phase of existence. Herbert Spencer's evolutionary philosophy confirmed this point by treating progression as a statistically irreversible trend underlying the multiplicity of natural activities at both the biological and psychological levels. Darwinism certainly sealed the fate of the attempt to see progress as the unfolding of a coherent plan with a single goal in man, and in principle it made the whole idea of progression obsolete. But in fact it was several decades before the majority of naturalists were able to shake themselves out of the belief that evolution manifested an overall direction that could in some sense be termed an advance.

Charles Darwin:
A New Basis for Progressionism

> When on board H.M.S. *Beagle*, as a naturalist, I was much struck with certain facts in the distribution of the inhabitants of South America, and in the geological relations of the present to the past inhabitants of that continent.

Thus Darwin began his introduction to the *Origin of Species*, making it clear that paleontological discoveries played some part in his conversion to evolutionism. In the Tertiary rocks of South America he had found numerous mammalian remains resembling modern forms such as the armadillo and the ground sloth, but much larger in size. From this evidence he formulated a "law of succession of types" in his *Journal of Researches*, and it is evident that from the first he had seen the significance of the many parallel lines of development in the fossil record, thereby escaping the fascination of the apparently simple progressive sequences which formed the basis of the *Vestiges* debate.[2] Owen, to whom the fossils were entrusted for description on the *Beagle's* return, later came to appreciate the same point, and even claimed to have anticipated Darwin in proposing the law.[3] But Darwin had gone much further in searching for a natural explanation of this phenomenon and of the strange geographical distribution of species in places such as the Galapagos Islands. An interest in animal breeding turned him to the study of variation and selection, which soon combined with a chance reading of Thomas Malthus' *Essay on the Principle of Population* to give the theory of natural selection. In any generation, he realized, those individuals best adapted to changing conditions will survive and procreate, while those least adapted will have fewer offspring and may die off altogether. Over a long period of time the species would thus change to a potentially unlimited extent. This was the theory which Darwin first sketched out in 1842, elaborated in his "Essay" of 1844 and finally published in the *Origin of Species* in 1859.

Natural selection was not immediately concerned with a species' level of organization, but there were ways in which it could have generated a progressionist interpretation of the history of life. Had Darwin accepted a directional theory of geology, for instance, he might have been led to such an interpretation by the same kind of logic as that accepted by the early catastrophists. At a late period in his life he did, in fact, express a brief interest in the theory of the declining carbon dioxide content of the atmosphere as revived by John Ball.[4] But from the beginning he had been attracted to Lyell's uniformitarianism— mostly, perhaps, because of its rejection of catastrophic extinction, but an inevitable consequence was a permanent lack of real interest in the possibility of directional influences on evolution from this source. Even the well-known debate over Lord Kelvin's support for the cooling- earth theory centered more on the lack of time this would allow for evolution to take place than on the possible influence of the cooling on the succession of forms. Any element of direction or progression in Darwin's theory would thus have to arise as a by-product of natural selection itself.

Darwin's initial difficulty in deciding if there could be any connection between natural selection and progression arose out of the fact that he was not even sure whether any precise meaning could be given to the concept of "highness." In a letter to J. D. Hooker in 1854 he completely rejected the often used comparison with man as a means of determining an organism's rank, suggesting instead that "highness" could perhaps best be defined in terms of the amount of deviation from the most generalized embryonic form of the class to which the species belongs.[5] This very limited concept illustrates how much Darwin's thinking on the issue was dictated by his early recognition of the divergence constantly taking place in the history of life. In his "Essay" of 1844 he had already noted that the ancient forms were "less widely divided" than those of today, and he soon realized that divergence of structure must be an inevitable consequence of natural selection.[6] This made the old man-centered hierarchy obsolete, and if he was to retain the idea of progression Darwin would have to grope his way toward an absolute scale of organization that could be applied to every part of a multilinear process. Even such an alternative would be difficult to define properly—hence Darwin's comment in his copy of *Vestiges*: "Never use word[s] higher and lower."[7] Nevertheless he could not escape the feeling that an experienced naturalist could sense the relative complexities of the organisms with which he dealt, thereby providing at least an intuitive scale of organization. The question then became: if such a redefinition of the hierarchy were possible, would natural selection produce a general tendency for organisms to mount up the scale, i.e., a progression?

Obviously there could be no necessary trend toward perfection in Darwin's theory—this was one reason why he objected when it was compared to that of Lamarck.[8] In the case of parasites in particular, increasing specialization for a certain way of life might lead to a positive degradation of structure. But in the "Essay" of 1844 he admitted that "we might expect to find some tendency toward progressive complication in the successive production of new forms."[9] And in a letter written to Hooker in 1858 he argued that over a long period of time the tendency of a species to improve its competitive ability "will make the organization higher in every sense of the word."[10] Any highly evolved form would have a greater survival capacity when measured against its own ancestors, or against other forms evolved in a more isolated place where competition was less vigorous. Thus Darwin believed that the present Eurasian flora and fauna would exterminate those of Australia where they were brought into contact, and with some misgivings he accepted that this "competitive highness" would correspond to an absolute difference in the level of organization.[11] This point was made quite specifically in the *Origin*:

> The inhabitants of each successive period in the world's history have beaten their predecessors in the race for life, and are, in so far, higher in the scale of nature; and this may account for that vague yet ill-defined sentiment, felt by so many paleontologists, that organization on the whole has progressed.[12]

Or, as Darwin wrote in the conclusion: "as natural selection works solely by and for the good of each being all corporeal and mental endowments will tend toward perfection."[13] While clearly rejecting the old linear progressionism, Darwin had thus followed the more advanced paleontologists such as Owen in accepting a more flexible definition of progression by which all forms could be said to have advanced within the context of their own kind of structure.

It is all the more necessary to stress that Darwin's kind of progression was quite different from the old linear variety because the two are still occasionally treated as equivalent, as in Maurice Mandelbaum's recent study of nineteenth century thought.[14] One particular area where confusion may be generated on this score is Darwin's attitude toward embryology, especially his acceptance of a recapitulation theory. Jane Oppenheimer's well-known study describes his commitment to the belief that ancestral forms must have resembled the embryos of their modern descendants and shows that he explicitly derived this from Agassiz.[15] But this relationship must be evaluated with some care, since if there were a significant connection here it might imply that Darwin still felt some attachment to the idea that a progressive direction has been programmed into evolution. (It may be significant in this respect that Ernst Haeckel's later "biogenetic law"— which closely resembled the old recapitulation theory—was associated with a rather un-Darwinian element of progressionism in his approach to evolution.[16]) But in fact it is clear that Darwin's understanding of recapitulation was totally different from Agassiz's. The Swiss naturalist had argued for the parallel between embryological development and the historical unfolding of life because it seemed to provide evidence that both processes were governed by the same divinely imposed plan. Darwin, on the other hand, showed in both his "Essay" of 1844 and the *Origin of Species* that recapitulation should occur as a natural result of adaptation by selection. His discussions of embryology concentrate on the fact that a closer resemblance exists between the embryonic than between the adult forms within any taxonomic group.[17] This, he held, would follow from natural selection's tendency to modify the original ancestral form by acting on the adult stages, leaving the earlier development of the embryo unchanged. For this reason there would be an apparent recapitulation of phylogeny by ontogeny, since the present embryos of the group would to some extent resemble the undifferentiated ancestral form. There was no question of either embryological or evolutionary development being predesigned.

It is particularly important to note that Darwin was never really attracted to the old law of parallelism between embryological development and the hierarchy of vertebrate classes. Agassiz had argued for this because it fitted in with his belief that the overall plan of vertebrate history was based on an ascent through the classes toward man. But in general Darwin confined his comments about embryology to developments occurring *within* the classes; almost the only exception to this is his speculation about the origin of the vertebrate type, and even here he did not imply that the embryo passes through the hierarchy of classes as displayed by the fossil record. As Oppenheimer points out, this speculation first occurs in the 1844 "Essay," where Darwin argued on embryological grounds that the fish must have been the first members of the type. The point was not mentioned in the first edition of the *Origin*, but reappeared in a modified form in the fourth edition, after Darwin had read with enthusiasm the embryological conclusions of Fritz Müller's *Für Darwin* (1864).

> So again it is probable, from what we know of the embryos of mammals, birds, fishes and reptiles, that these animals are the modified descendants

of some ancient progenitor, which was furnished in its adult state with branchiae, a swim-bladder, four fin-like limbs and a long tail, all fitted for an aquatic life.[18]

Note that at this later stage, Darwin carefully avoids implying that the original form was a fish in anything like the modern state of the class. Essentially all he wished to show was that the fishes as a whole are less differentiated from the original embryonic form of all vertebrates, and hence that the earliest members of the type would resemble this class more than any other. The emphasis is on the more general structure of both embryonic and ancestral forms, not on the actual hierarchy of the classes. Clearly, Darwin did accept a kind of recapitulation theory in which embryos to some extent resemble ancestral forms, and in this he differed from von Baer who repudiated this possibility altogether. But this does not imply that he abandoned von Baer's new conception of embryological growth for the old law of parallelism, where the embryo passes through the succession of classes. In effect—as E. S. Russell pointed out as long ago as 1916[19]—he was working with a new version of the recapitulation theory actually based on von Baer's embryology, a possibility that von Baer himself had failed to appreciate but which had already been brought out by W.B. Carpenter in 1851.[20] The fact that at first Darwin acknowledged Agassiz's work in connection with embryology, not von Baer's, signifies only that the former had openly supported recapitulation (if in an out-of-date form) while the latter had repudiated it. Indeed by 1857 even Agassiz was aware of von Baer's work and had given up the parallelism between the classes.[21] Darwin *was* interested in the possibility that embryos resemble ancestors, but only within the context of the new embryology and his own belief that natural selection brought about a gradual specialization of structure. His acceptance of this new and quite different kind of recapitulation theory thus confirms the gulf between his approach to progression and the old linear version originally proposed by Agassiz. For Darwin the evolutionary process was not designed; both progression and recapitulation were merely by-products of natural selection's constant tendency to produce divergence and specialization of structure.

Two chapters of the *Origin* were devoted to paleontology, the first using "the imperfection of the geological record" to explain the still troublesome gaps in the history of life, the second looking more positively for trends that would serve as evidence for the theory. The extremes of discontinuous progression were no longer plausible, and Darwin did not bother to defend himself against this mode of attack. But there were still some notable gaps in the record—particularly the apparently sudden development of the bony fishes in the Cretaceous.[22] Darwin argued that, considering the hazards of fossilization, it was hardly surprising that at least some major developments will have occurred without leaving any trace, especially when the divergence took place very rapidly as a class expanded into a whole new range of ecological possibilities. The very beginning of the record itself presented a problem, however, since the multitude of Cambrian forms appeared without any sign of a previous development. Darwin held that all the different forms of life have evolved from at the most a handful of original types, and he was forced to admit that the lack of evidence for this first stage of the process was very puzzling. The only excuse he could offer was that many of the most primitive forms at the beginning of life would have had no solid structure to leave behind in the rocks.[23] A glimmering of hope appeared a little later with the discovery of *Eozoön canadense*, a structure apparently resembling a large variety of Foraminifera unearthed by Sir William Logan in the ancient rocks of Canada.[24] The organic origin of these structures was at first strongly upheld by Sir John W. Dawson of Montreal and by W.B. Carpenter, who had become an expert on the Foraminifera. Darwin greeted the discovery with some enthusiasm in his fourth edition (1866) and hoped that the future would reveal a whole range of Precambrian fossils.[25] Eventually, however, opinion shifted to regarding the structures as the result of purely mineral activity, leaving the evolutionists exactly where they had begun. Even today only a limited number of rather ambiguous organic remains have been identified from these early rocks.

In his next chapter "On the geological succession of organic beings," Darwin tried to show that once the incomplete nature of the record was taken into account, its general outlines were in agreement with his theory. Inevitably, he concentrated mainly on divergence and specialization, not progression. The pattern of development to be expected from natural selection had already been illustrated by the well-known diagram, part of which is reproduced here. (See Plate XI.) Each line of development has a constant tendency to diverge into multiple lines which then move further away from one another, the multiplication of species being compensated for by the occasional extinction of a line caused when a form can no longer adapt quickly enough to keep up with changing conditions. Darwin refused, however, to give concrete examples of this pattern in the form of hypothetical genealogies for the present species—reading *Vestiges* and its critics had shown him that there was far too little evidence available to do this safely.[26] He pointed out that given a pair of related modern forms, the common ancestor would not be a simple intermediate and identification would be very difficult unless a whole range of linking forms was discovered, a most unlikely possibility considering the state of the record. But Darwin did feel that on the whole the distribution of fossil forms indicated trends of divergence that were well in accordance with his theory. This was where Owen's work came in very useful: Darwin cited his demonstration that the ruminants and pachyderms (the two most distinct mammalian orders, according to Cuvier) had been linked together by fossils from the early Tertiary that were intermediate in character between them.[27] Although Owen refused to accept natural selection as an explanation of such trends, the evidence he provided was quite consistent with Darwin's interpretation.

The irregularities to be expected from the operations of natural selection would themselves render the fossil record difficult to interpret, creating exceptions that superficially might seem to contradict the theory. A case in point was the very limited amount of change exhibited in some areas, as when certain primitive forms persisted almost unchanged through several geological periods. Darwin emphasized that if a form becomes well adapted to a very stable environment it will persist unchanged—natural selection gives no necessarily progressive trends.[28] In later editions of the *Origin* he illustrated this by referring to W.B. Carpenter's objection that the fossil Foraminifera showed no sign of an advance toward a higher type.[29] Since these simple organisms do, in fact, inhabit a stable environment, their lack of change was quite in accordance with the theory. In a somewhat more complex case, Hugh Falconer's work on fossil elephants had revealed that the periods of existence of the various forms did not correspond to any natural classification; i.e., the more highly evolved forms sometimes appeared first in the record.[30] This could result when an ancient species lasted unchanged for a long period of time, but only left fossil remains toward the end of its period after related forms had undergone further evolution and divergence. Darwin's own law of the succession of types itself posed certain problems: it was hardly feasible, for instance, that the modern South American forms could have evolved from their giant relatives that had been discovered in the Tertiary rocks. Darwin argued that the giant species have died out completely, but there must have been similar forms of a more reasonable size which served as the ancestors of those of today.[31] Often only one species would have survived from a particular fossil genus, to become, through divergence, the parent of a new genus containing a

number of species. Unfortunately, the hazards of fossilization will
ensure that the paleontologist will rarely discover the ancestor itself,
although he might have access to one of the extinct members of the
ancient genus. At varying levels of importance it was to become
increasingly obvious that the fossil record is littered with forms that
were related to the ancestors of the modern animals, but were not on
the direct line of descent.

Although Darwin refused to speculate about the ancestry of the
modern species in print, a letter written to Lyell in 1860 gives a good
idea of his expectations about the fossil record and the nature of
biological progress. He used the diagrams reproduced here to illustrate
what he considered to be the two possible modes by which the true
mammals could have evolved. (See Plate XII.) On the basis of the
earlier linear approach to progression, many workers might have
assumed that marsupial-like creatures were the ancestors of the later
placentals, as in diagram II, and this was supported by the normal
interpretation of the fossil record, i.e., the identification of some of the
Secondary mammals as "marsupials." But Darwin expressed a
preference for the scheme represented in the first diagram, since this
fitted in with his ideas on the production of the modern forms through
divergence. He admitted that the second version might be shown to be
more probable if it could be demonstrated that the embryos of
placentals resemble adult marsupials, but he was instinctively
suspicious of a system which held that forms could still be alive which
resembled the ancient ancestors of the class. In effect, Darwin did not
want to see the development as a simple progression measured in terms
of the hierarchy that could be recognized within the modern forms. He
preferred to think of the marsupials as a distinct line of specialization,
a line which may have differentiated itself to a lesser extent from the
more generalized ancestral forms, but which was by no means a direct
continuation of an almost identical structure. There is no doubt that
Darwin held the true mammals to be higher than the marsupials in
terms of competitive ability, and he felt that this would imply that they
were higher in the sense of being structurally more complex. But he
would probably have felt that they were both in some degree superior
to the undifferentiated ancestral forms, although natural selection
would have acted less forcefully on the marsupials because of their
more isolated environment. Thus the process represented by the
diagram could still be seen as a progress, but since it was accompanied
by constant divergence it was quite different from the old idea of an
ascent through a more or less linear hierarchy.

Darwin's ideas on the overall structure of the evolutionary process were echoed by Alfred Russel Wallace, the co-discoverer of natural selection. H.L. McKinney has shown that in his early species notebooks Wallace already accepted progression but saw that advances *within* the classes need not always be related to the general advance *between* the classes.[32] He thought that the first reptiles might arise from lower fishes, both classes then developing in separate directions. In his first paper on evolution in 1855 he argued for an irregular progression and related this to the idea that the history of life is a branching process that could be represented as a structure resembling a tree.[33] When he finally worked out the mechanism of natural selection, his first paper clearly implied that this would tend in the long run to produce forms that were higher in every sense of the word.[34] He also shared Darwin's belief that this would have to be seen against the background of constant divergence and specialization already demonstrated in the fossil record by Owen.[35] In a review of the 1868 edition of Lyell's *Principles of Geology* (which finally accepted evolution) he argued that although the fossil record was far from complete, it was most unlikely that the basic trends it exhibited were illusory. Since we actually observe a convergence backwards toward simpler and more general types, it was legitimate to infer that the whole development of life must have arisen from a small number of very primitive forms.[36] This position was maintained throughout the rest of Wallace's career[37] and only on one point did he seriously disagree with Darwin. This was on the nature of human evolution, which he soon decided to have been a non-utilitarian process that must have been shaped by direct divine supervision.[38] The fact that he abandoned natural selection as the driving force only in the last stages shows, however, the difference between the new progressionism and the old. Man could no longer be seen as the inevitable goal of the whole creation, since progress was just a general trend that has shaped all of the various evolutionary lines. It was merely the case that one of these lines had eventually reached a particularly promising point where it could be singled out for special treatment in its last phases of development.

If Wallace's lack of confidence in natural selection was confined to the last stages of human evolution, many others were at first skeptical about the mechanism's validity at all stages of the evolutionary process. Alvar Ellegård's survey of the Darwinian debate reveals that in the decade after the *Origin* was published there was a definite swing toward evolution, but not toward Darwin's explanation of it.[39] This in itself suggests that the earlier arguments based on the extreme discontinuity of the fossil record had lost their force, but shows that many workers preferred to turn to Owen's belief that the development of life is under divine control at all times. Yet it was Owen himself who had helped to demonstrate the lack of any coherent structure in the fossil record, replacing Agassiz's view of progression toward man with a new approach compatible with the principle of divergence and specialization. This interpretation made it almost impossible to believe that there was a central theme in the history of life and at the same time threw doubts on the earlier argument that the perfect adaptation of all living forms to their environment is a sign of benevolent design. Writers such as W.B. Carpenter, St. George Mivart, and the Duke of Argyll tried their best to adapt the idea of divine control to the increasingly obvious irregularity of the evolutionary process, but in the long run it was the Darwinian approach that fitted in more easily with the new interpretation of the fossil record and the later nineteenth century saw a steady decline in the popularity of theistic evolutionism.[40]

The spread of the Darwinian viewpoint was almost certainly encouraged by the growing influence of Herbert Spencer's "Synthetic Philosophy." This provided biological evolutionism with a complete intellectual background in which the new indirect kind of progressionism figured prominently as a basic universal trend. Spencer himself distrusted the paleontological evidence for organic progress, suggesting that the fossil record might represent only the process by which the present continents were stocked by forms migrating in from elsewhere.[41] Although he wrote in favor of transmutation as early as 1852, he could not accept that the process was governed by a preordained plan of advance and preferred to see adaptation to changing conditions as the leading factor involved.[42] Having been impressed by W.B. Carpenter's account of von Baer's embryology, he elaborated a philosophy in the course of the 1850s which took the process of development from homogeneity to heterogeneity of structure as fundamental to all natural activities. Elaborated in his *First Principles* of 1860–62, the philosophy was applied in detail to the evolution of life in the *Principles of Biology*, issued in 1863–64. The basic principle of development indicated that in response to the demands of adaptation there would be a constant diversification of organic structure; that is, a multiplication of the number of species, accompanied by a statistically inevitable increase in the level of organization of the whole ensemble.[43] Spencer preferred the inheritance of acquired characteristics to natural selection, but his philosophy was easily (if somewhat unfairly) connected with the Darwinian theory because of his emphasis on the *laissez-faire* approach to social issues.[44] Indeed it was almost certainly his works that popularized the name "evolution" in its modern context and gave the word a distinctly progressionist flavor.[45] Biological evolutionism became just one manifestation of the universal law of progress, but at the same time it was emphasized that the advance of life was a general trend affecting all the lines of development produced by the requirements of adaptation, not the result of a specific plan. In Spencer's philosophy man was the head of creation only in the sense that the evolutionary process had succeeded in pushing him further than any of the other lines, into a new phase of development where intellectual and social progress became important. Darwin himself tried to explain the production of man and his societies as a consequence of the general workings of natural selection in his *Descent of Man* of 1871. In the popular mind at least, faith in the universal tendency toward progress could replace the old dependence on direct divine control. By integrating man thoroughly into nature he could retain his place at the head of creation, not as the key to an artificially structured divine plan but as the highest point so far reached by a general tendency that has affected all branches of living development. Nor can the influence of Spencer's philosophy on the scientists themselves be ignored; it is significant that as late as 1907 R.H. Lock's *Variety, Heredity and Evolution*—a work devoted solely to the technical problems raised by Darwinism—still began with a progressionist definition of evolution derived from Spencer.[46]

Darwinian Paleontology and Progression

Although Darwin himself was reluctant to attempt a detailed reconstruction of the history of life, it is perhaps not surprising that some of his followers were less cautious. To be sure, there were enormous difficulties presented by the evidently incomplete state of the fossil record, and from time to time voices were raised in protest against the sometimes over-hasty attempts that were made. In 1880 Louis Agassiz's son Alexander—unlike his father a reluctant convert to evolution—argued that the proposed genealogies were "built on air" and urged paleontologists to devote their efforts to more fruitful lines of research.[47] But even Darwin thought that Agassiz overemphasized the difficulties[48] and in any case the process of reconstruction was well underway by this time. Where the fossil evidence failed, embryology and comparative anatomy were used to provide hypothetical links between the known forms. This was scientifically respectable in the sense that it could always be hoped that such speculations would eventually be tested by the discovery of new fossil specimens. To an unfortunately large extent this has proved impossible, so that modern paleontologists still grumble about the imperfection of the record in the same pessimistic tone as Darwin himself. Yet a few important clues were discovered, providing hints about some of the great mysteries such as the origins of the vertebrate classes. Much was still left to the imagination, but the steady advance of paleontological discovery offered at least the hope of showing that the evolutionists' speculations were running along the right lines.

Almost inevitably these researches sounded the death knell of the linear view of development, confirming once and for all that the fossil record could only be understood as the remnant of a highly complex evolutionary process. Perhaps the only major paleontologist of this later generation to start out from a partly linear view of development was the American Edward Drinker Cope, whose early paper "On the origin of genera" (1868) suggested that the generic forms represent fixed and divinely preordained hierarchies along which a series of different lines corresponding to the various species advance.[49] This closely resembled Chambers' earlier idea of "stirpes" and it is significant that Cope too held that the plan of advance was related to the development of the individual embryo. As his paleontological studies expanded, however, he rapidly abandoned this idea in favor of a Lamarckian theory of adaptation in which the inheritance of acquired characteristics became the chief driving force of evolution. Thus, although he refused to accept natural selection, Cope's superficial conception of the history of life became close to that of the Darwinists and he was able to make notable advances in the field of reconstructing the hidden evolutionary steps.[50] The connection between embryological development and evolution was also supported by Cope's colleague Alpheus Hyatt, who proposed a law of the acceleration of growth to account for the parallels he observed between the fossil history of the ammonites and what he had discovered about the early growth of the individual shells.[51] Here was apparently more solid evidence for what came to be known as "orthogenesis"— unusually linear sequences of development with no obvious adaptive significance. The Darwinists were often at a loss to account for this phenomenon, but it had little resemblance to the old linear progressionism since the sequences that were established confined themselves to particular groups and did not link together to form any kind of overall advance of life. They were minor anomalies that had little power to undermine the Darwinists' conviction that the real essence of the evolutionary process was its continuous divergence. It was occasionally admitted even by the Darwinists themselves that several evolutionary lines might move in parallel toward the same general goal—Ernst Haeckel, for instance, argued that the placental mammals may have originated from a number of marsupial groups advancing toward the new level of organization.[52] An opponent of Darwin such as St. George Mivart might use this to argue that life was designed with an inherent predisposition to advance in a certain direction, but the Darwinists themselves felt that there was nothing unnatural about selection driving a number of as yet unspecialized forms toward a particularly advantageous new level of organization. This interpretation of the origin of the mammals is, in fact, still favored by modern paleontologists, although they would follow Darwin in assuming that the marsupials are not a direct relic of the process. The problem was (and still is) to find fossil evidence of the steps by which one or more lines within an earlier class advanced toward the new grade of organization. What was apparent to all, however, was the obvious fact that when the conditions became favorable, the new class would blossom out into a host of adaptive specializations, as did the mammals during the early Tertiary.

A description of the process by which the later nineteenth century naturalists laid the foundations of our modern knowledge of the development of life would take a book in itself, much of which would be irrelevant to our general theme of progressionism. The following section will study just two workers who made notable contributions to the field as examples: O.C. Marsh and T.H. Huxley. Both were staunch Darwinians who collaborated on a number of occasions, and both used the theory as the foundation of their attempted reconstructions. But they adopted opposing views on the relationship of Darwinism to progressionism. Marsh followed what was at first the more popular line, concluding that the fossil record was fully in agreement with Darwin's assumption that in the long run natural selection must give rise to progress. His "law of brain growth" may be counted as a paleontological equivalent of the more philosophical progressionism of writers such as Spencer and Haeckel. Huxley, on the other hand, was suspicious of the fossil evidence for progression, holding that the concept was almost incapable of meaningful definition and that it was pointless to connect the new evolutionary theory with such an outdated notion. The contributions of these two naturalists provide an excellent illustration of how the new breed of paleontologist tried to come to grips with the complexities of the evolutionary process, while at the same time showing the two extremes of opinion on whether there was an underlying tendency toward progress. Marsh represents the typical late nineteenth century attachment to the belief that the universe is at bottom a progressive system (even if the details of the advance are not predesigned by the Creator). But Huxley points the way toward the increasing suspicion of biological progressionism that has grown up in the twentieth century as the earlier period's optimistic faith in general progress has crumbled.

One of the most well-known areas in which Huxley and Marsh cooperated was in the study of the evolution of the horse. Owen had established the general rule that in structures such as their teeth and hooves, the Tertiary mammals became progressively more specialized in the course of the period's history. But there were no continuous sequences of fossils that would illustrate this process happening in a real line of evolutionary development. To establish such a line would give a great boost to the evolutionists' morale, and in the course of the 1860s Huxley made extensive studies of the European fossil horses in the hope of tracing the modern one-toed species back through a continuous series of fossils to a more generalized multi-toed form. He was able to connect the modern horse back to the three-toed *Anchiotherium* and *Plagiohippus* of the Eocene.[53] But when in 1876 he went to visit Marsh in America he found that the latter had already established a far more convincing series that linked the modern horse to the four-toed *Orohippus*.[54] (See Plate XIII.) In his *American Addresses* Huxley announced his conversion to the American origin of the horse and hailed Marsh's work as "demonstrative evidence of evolution."[55] In a footnote added to this work he was also able to comment on Marsh's discovery of the final link in the chain; the five-toed *Eohippus* of the lowest Eocene rocks.[56] The significance of this sequence has sometimes been questioned, but Huxley and Marsh had no doubt that they had uncovered not only a continuous evolutionary series but also one which illustrated the basic mechanism of the Darwinian theory, specialization resulting from the continued pressure of adaptation. Since the line leading to the horse was only one out of a multitude that would have to exist in order to account for the host of modern mammals, it was now more than ever clear that the old linear concept of development was unrealistic.

Further attempts to understand the evolution of the mammals led to an increased sense of the scope of the divergence that the class has undergone. Marsh's researches uncovered a whole new ungulate order which he named Dinocerata, and in his monograph on these forms he speculated on the origin of the ungulates as a whole and of the mammals themselves. He gave the by now familiar tree-like diagram of evolutionary branching, putting the hypothetical ungulate ancestor, "Protoungulata," in the mid-Cretaceous.[57] He was now convinced that the basic differentiation of the group from the rest of the mammals must have occurred before the Tertiary and guessed that the original ungulates were derived from more generalized mammalian forms probably lying as far back as the Permian. In this he was following a suggestion already made by Huxley, who had also come to appreciate that the diversification of the class was a far more extensive and slow-moving process than had at first been anticipated. This was confirmed later in the century by workers such as C. Depéret and K. von Zittel.[58] They showed that even the mammalian orders are polytypic; i.e., are made up of genera that have been developing separately for a long period of time—a sure sign that the original production of the orders themselves must have occurred very far back in the past. Earlier workers had known that the mammals appeared long before the Tertiary, but the lack of fossil evidence from the Cretaceous had prevented them from appreciating that the class had already begun to differentiate itself into its major subdivisions during this earlier period. Through a vast period of geological time the mammals had advanced on a number of different fronts, just like the other classes, a point which stressed the extreme non-linearity of the evolutionary process and the impossibility of treating the history of the class as an advance toward man.

The question of the actual origin of the classes was one of the most pressing that faced the evolutionary paleontologists. It was becoming increasingly evident that the classes went much further back in time than had at first been imagined—in some respects Lyell had been surprisingly close to the true position with his old non-progressionism. But as yet there was little information about and no real proof of the evolution of one class from another during the early stages of the record. Working on anatomical grounds, Huxley tried to assess the affinities between the classes and hence their degree of evolutionary relationship. He urged the separation of the amphibians from the reptiles and their inclusion along with the fishes in a larger group to be called the Ichthyopsida. He also associated the known reptiles with the birds in another group the Sauropsida. This approach stressed the differences between the then-known reptiles and the mammals and contradicted Owen's claim that the dinosaurs had certain resemblances to the mammals. Huxley now suggested that the dinosaurs' only real connection could be with the birds, i.e., the birds could have evolved from forms related to the dinosaurs, but not the mammals. Indeed, Huxley was so keenly aware of this point that he suggested the possibility of the mammals evolving not from the reptiles but directly from the amphibians, via some as yet unknown intermediate form.[59] This idea is no longer accepted, but at first there was no evidence to gainsay it and some new discoveries were even interpreted as favorable to the theory.[60] Only toward the end of the century did the Karoo formation of South Africa begin to reveal the existence of a series of what came to be known as the "mammal-like reptiles," described by workers such as Harry Govier Seeley.[61] We now regard the mammals as having evolved via these forms from certain true reptiles, the pelycosaurs of the Carboniferous. Huxley had exaggerated the problem, but he had been quite right to emphasize that the fossil reptiles known in the 1860s could be associated with the birds and not the mammals. He had correctly predicted that the paleontologists would have to come up with some totally new forms to bridge the reptile-mammal gap, and that these would at the very least reveal the existence of another line of reptilian development running parallel to those already known through a vast period of time. Again the multilinearity of evolutionary development was making itself apparent.

The discoveries that confirmed the existence of a new line of reptiles leading toward the mammals were not made until the end of the century, and to begin with Huxley and Marsh had to concentrate on the evidence linking the known reptiles to the birds. In 1868 Huxley published a paper "On the animals which are most nearly intermediate between birds and reptiles." He concentrated on a reptile, *Compsognathus longipes* from the Trias, which resembled a small dinosaur except that its hind limbs and feet were so similar to those of a bird that it deserved to be called a veritable "missing link."[62] (It should be noted that it was not until much later that H.G. Seeley proposed the modern arrangement of dividing the dinosaurs into two orders—the Saurischia and Ornithischia—of which only the latter resemble the birds in the structure of the pelvis.[63]) Huxley was cautious enough to admit that *Compsognathus* was probably not on the direct line linking the classes; in fact he suspected that the birds were already in existence at this time with *Compsognathus* being only a less thoroughly modified version of the ancestral form. This again was an overestimation of the age of a class, although Huxley was quite right to assume that the point of divergence was a long way back in time. Modern paleontologists believe that the dinosaurs and birds are quite distinct, having developed in parallel from some much older ancestral form.

In his 1868 paper Huxley also mentioned another possible intermediate, *Archaeopterix*, a form much more closely allied to the birds from the Jurassic. Owen had described a specimen lacking only the head in 1863, concluding that it was clearly a bird, although with certain highly generalized characteristics.[64] Huxley criticized Owen's description, but agreed that *Archaeopterix* would have to be regarded as a bird even if a head with teeth were discovered.[65] As a matter of fact such a head was found in 1877, but even before this Huxley had begun to lay more stress on this form as a real intermediate between the two classes. It was presented in this light in his *American Addresses*, along with the toothed birds from the Cretaceous which had now been discovered by Marsh.[66] Two such forms had been unearthed in an exploration of Kansas in 1872; Marsh named them *Ichthyornis dispar* and *Hesperornis regalis* (See Plate XIV.) and soon created a new sub-class for them, the Odontornithes.[67] He expressed the hope that birds of an even more reptilian character would be found in the Jurassic, but followed Huxley in speculating that the origin of the class would lie even further back in time.[68] Both workers agreed that the different forms of toothed birds offered very strong evidence in favor of the evolutionists' claim that the classes must have emerged from one another. But they both appreciated the extent of the gaps that still remained and had a healthy, perhaps even an exaggerated, respect for the enormous amounts of time involved and the immense complexity and diversity of the evolutionary process.

Important as it was to provide evidence in favor of continuous development, the real significance of the studies described above lies in the revolution they produced in the naturalists' ideas on the structure of the evolutionary process. Every step that they made took the paleontologists further away not just from the old idea of linear development but also from the whole hierarchical conception of natural arrangements. The relationships between the classes, especially the connection of the mammals and birds with the reptiles, showed that there was no point at all in treating the history of the vertebrates as a process governed by a unified theme of development. It was now clear that the two "higher" classes have arisen separately out of quite different lines of reptilian evolution, making it impossible to follow the earlier progressionists in their belief that the birds were a step on the way toward the mammals. Here was the most powerful evidence of all to show that the development of life was not aimed principally at the appearance of man, and that the human form was not the key to the understanding of the whole vertebrate type. Even within a single class it was now confirmed that there was nothing that could correspond to a simple progression. The early division of all classes, including the mammals, into multiple lines of development made it virtually impossible to construct a progression out of the sequence in which the orders and families appeared. Any attempt even to define a hierarchy of organization within a class was complicated by the fact that all too often forms could be seen to have advanced in some parts of their structure but not in others. As Marsh pointed out in the case of his toothed birds, *Hesperornis* had low, generalized teeth but a very advanced vertebral structure, while *Ichthyornis* had acquired highly specialized teeth coupled with more primitive vertebrae. He commented that this kind of irregular development was a "fundamental principle of evolution."[69] Owen, of course, had recognized this phenomenon even before the *Origin* was published, especially in connection with the fishes, but in general it had been ignored by the earlier paleontologists who had been determined to argue for or against a more or less linear view of development. Now the fact that natural forms could not be twisted into an unambiguous hierarchy against which to measure progression was seen to be a fundamental consequence of the nature of evolutionary development. To Marsh and the other Darwinian paleontologists it seemed clearly impossible that the totally irregular kind of development revealed by the fossil record could be governed by a preordained and unified plan, least of all one aimed at the production of a particular goal such as man. Darwin's theory was really the only one that could cope with such a phenomenon.

Darwin himself had realized the futility of all hierarchical arrangements even without the stimulus of these paleontological developments and had rejected the corresponding notion of progression. But as we have seen he still felt that the naturalist had the ability roughly to gauge the "highness" of an organism, and rather than abandon progressionism altogether he had suggested that natural selection would in the long run push all forms of life toward higher levels of organization. The problem now was to find a reasonably objective method of confirming this. Most naturalists would have agreed that the birds and mammals, for instance, were higher than the reptiles, but since they were parallel developments it was evident that some means of measuring the progression would have to be used that was independent of their particular structures. One of the most promising attempts to solve the problem was made by Marsh, with his "law of brain growth." The brain was a useful organ to choose, since changes in its size could be measured fairly objectively, while increasing brain size would be taken by the majority of naturalists as a sign of general progression in the level of organization. By surveying the various lines of development in the fossil record, Marsh felt that he could indeed show that evolution has produced a general advance as measured by this criterion. The law was first announced in 1874 and then stated fully in 1876; it comprised the following propositions:

> *First.* All Tertiary Mammals had small brains. *Second.* There was a gradual increase in the size of the brain during this period. *Third.* This increase was confined mainly to the cerebral hemispheres, or higher portions of the brain. *Fourth.* In some groups the convolutions of the brain have gradually become more complicated. *Fifth.* In some the cerebellum and the olefactory lobes have even diminished in size. There is some evidence that the same general law of brain growth holds good for the Birds and Reptiles, from the Cretaceous to the present time.[70]

In his *Dinocerata* monograph Marsh repeated the law and added that further research showed it to hold good for the birds and reptiles as far back as the Jurassic.[71] In an earlier time a law of this generality would probably have formed part of a theory of necessary progressive development. But Marsh made it clear that it could be understood in a purely Darwinian framework: brain size tends to increase simply because extra intelligence is an important weapon in the struggle for existence, but there is no automatic guarantee of progress. Thus he declared in addition that:

> (1) The brain of a mammal belonging to a vigorous race, fitted for long survival, is larger than the average brain of that period, in the same group.
> (2) The brain of a mammal of a declining race is smaller than the average of its contemporaries of the same group.[72]

Some groups with insufficient intelligence might decline, but the fossil record showed that in the long run the tendency was toward higher intelligence. Putting it another way, "In the long struggle for existence during Tertiary times, the big brains won, then as now."[73] The law of brain growth thus appeared to offer an ideal way of objectifying the new sense of progression which arose out of Darwin's own feelings about natural selection and Spencer's wider philosophy of universal complexification.

Had there been any suitable human fossils available, Marsh would probably have tried to apply his law to them in an effort to clarify the way in which man's brain can be said to represent the high point yet achieved by the evolutionary process. But as Huxley had shown in his *Man's Place in Nature* of 1863, the available remains of fossil man—the Engis and Neanderthal skulls—were totally unsatisfactory as guides to man's ancestry. The link between the human species and the rest of the animal kingdom had to be demonstrated on anatomical grounds, hence Huxley's debate with Owen. Only at the very end of the century were the speculations of Haeckel and others about the course of human evolution crowned by the discovery of *Pithecanthropus erectus* (Java man), the first fossil that could reasonably be regarded as a link between man and his apelike ancestors.[74] It was in his study of man that Huxley was led to produce an almost Spencerian passage on the subject of progress, referring to "Nature's great progression from the formless to the formed—from the inorganic to the organic—from blind force to conscious intellect and will."[75] But if he could wax lyrical over the evolution of man and its significance, Huxley was highly suspicious of Darwin's progressionist interpretation of the new evolutionism. Where Marsh tried to provide objective paleontological evidence for the Darwinian version of progression, Huxley openly maintained that the record illustrated the lack of connection between evolution and increasing levels of organization. His views thus mark the beginning of the modern attitude whereby evolutionists simply ignore the whole issue of progression.

Already in an 1862 presidential address to the Geological Society, Huxley had expressed his suspicions of the fossil evidence for progression. After criticising the geologists' automatic assumption that similar formations in different parts of the world must have been laid down at the same time,[76] he went on to say that he was constantly amazed at the *lack* of significant difference between fossil and living forms. He pointed out that J.D. Hooker had shown that the paleobotanists had not found it necessary to create a single new order of plants beyond those needed to classify the living species.[77] Less than ten percent of the fossil animals required the creation of new orders. All paleontologists were aware, said Huxley, of the phenomenon of the "persistence of type"—the continuation of essentially similar forms over vast periods of geological time.[78] Once this was seen as a basic element of evolutionary development, it became obvious that there could be no necessarily progressive trend operating.

> What then does an impartial survey of the positively ascertained truths of paleontology testify in relation to the common doctrines of progressive modification, which suppose that modification to have taken place by a necessary progress from more to less embryonic forms, or from more to

less generalized types, within the limits of the period represented by the fossiliferous rocks?

It negatives these doctrines; for it either shows us no evidence of any such modifications, or demonstrates it to have been very slight; and as to the nature of that modification, it yields no evidence whatsoever that the earlier members of any long continued group were more generalized in structure than the later ones. To a certain extent, indeed, it may be said that imperfect ossification of the vertebral column is an embryonic character; but, on the other hand, it would be extremely incorrect to suppose that the vertebral columns of the older Vertebrata are in any sense embryonic in their whole structure.[79]

As Huxley later admitted, his own interest in evolution encouraged him to hope that the fossil record would at least reveal a trend toward increasing specialization of structure, but at this point in time he seems to have been suspicious of Owen's earlier attempts in this direction. In another presidential address in 1870, he retracted his opposition on the grounds that satisfactory evidence of such a trend had now been discovered. This included his own work on the horse and the elucidation of intermediate forms between the birds and reptiles.[80] But he still maintained his insistence on the importance of the permanence of type and asserted that, especially with the lower vertebrates, the modern forms were in no way superior to their ancestors. Thus while accepting that the fossil record did indicate a process of specialization, he refused to accept that there was any overall progressive trend. In fact, all of Huxley's later work in reconstructing the history of life was done without reference to progressionism—he does not, for instance, seem to have expressed any interest in Marsh's law of brain growth. His attitude illustrates a desire to work purely with the fundamental principles of Darwinism, avoiding the confusion of issues that would arise if the new theory were to be associated even indirectly with the old concepts of development. Evolution worked by adaptation and specialization; attempts to describe a clear-cut progression of the kind understood by earlier workers were doomed to failure, and it seemed to Huxley that it was better simply to ignore the whole concept in order to emphasize the superiority of the new approach in dealing with the complexities of the fossil record.

In the decades after 1860 Huxley's refusal to compromise with progressionism was by no means typical. The growing popularity of Spencer's philosophy—indeed the whole spirit of the age—ensured that in general "evolution" would be associated with progress. Marsh's law of brain growth does not seem untypical of a period when an

attempt to establish progression on a new footing was almost a
psychological necessity created by the demise of the old view that
development was designed by the Deity. The Darwinian view of progress
was radically different from the old concept, but it still served the
purpose of demonstrating man's position at the head of nature and of
pointing the way to further advances. In the early twentieth century,
however, Spencer's philosophy of universal progress lost its popularity, and even Darwinism itself came temporarily under fire from the
supporters of the new genetics. Alternatives such as the "mutation
theory" of Hugo de Vries did not challenge the basic outline of the
history of life established by the Darwinians (although they made the
element of absolute continuity unnecessary), and the same is true of the
Lamarckianism revived by E.D. Cope at the end of the nineteenth
century and occasionally supported by more recent paleontologists.
But in general evolutionists of all backgrounds came to appreciate the
wisdom of Huxley's decision to abandon once and for all the relic of
the old progressionism retained by some of the early Darwinians. The
rise of the "modern synthesis" between natural selection and genetics
has confirmed this trend—evolution has become a more complex
phenomenon than even Darwin imagined, but the lack of similarity to
the old progressionism is more striking than ever. Even the occasionally advocated theory of "orthogenesis"—based on the apparently non-
adaptive trends which seem to direct some areas of the fossil
sequence—has seldom been connected with progression, and in any
case the trends are not common enough to have convinced the majority
of paleontologists that a natural explanation is impossible.[81] Certainly,
the modern paleontologist will sometimes describe a form as
"primitive," but he generally has a fairly objective way of using the
term, for instance to describe a resemblance to some earlier form.
Efforts to trace out an advance in organization of the kind understood
in earlier times are few and far between, as are attempts to derive a
progressive trend from modern evolutionary theory. There is, perhaps,
a general feeling that the later forms of life are more advanced in the
sense that they are more independent of their environment, and
workers such as Julian Huxley, G.G. Simpson, and J.M. Thoday
have occasionally used this kind of criterion as a means of defining a
new kind of progress.[82] But such attempts are rare, they are quite
distinct from the normal approach adopted by the modern naturalist,
and they involve a reinterpretation of progression even more radical
than that implied by Marsh's law of brain growth. Julian Huxley, for
instance, admitted that most of his colleagues would regard his

speculations with distrust, while his own work in general makes it clear just how little resemblance there is between the new evolutionism and the old progressionism. The popularity of Teilhard de Chardin's writings may indicate a continuing need at some levels for reassurance that the universe does have a meaningful direction, but in general the scientists remain unconvinced.[83]

The middle and late nineteenth century were thus the high points of popularity for biological progressionism, despite a considerable shift of emphasis in the middle of the period. The basic idea that life has advanced toward higher forms in the course of its history first began to be seriously considered at the beginning of the century when a certain amount of rather discontinuous fossil evidence began to emerge. But at first its significance was limited by the fact that it was regarded as a simple consequence of the prevailing directional view of the earth's physical development. It was Agassiz who elevated progression to a new level of importance by making it a key factor in his vision of a succession of species linked by a transcendental divine plan centered ultimately on the production of man, while Chambers rapidly adapted the concept to the radically new view of natural development brought about by a continuously operating force. The fact that he still felt this force to be directed by a divine plan was largely ignored by those who could not accept that man was merely the last product of such a trend, and for some time the incompleteness of the evidence still favored those who maintained the absolute fixity of species. Gradually some of the major gaps were filled in, however, and the idea of continuous development began to be more widely accepted. But at the same time the increasing complexity of the fossil record forced a reassessment of the concept of progress, so that even before Darwin published the *Origin of Species* some workers were beginning to treat it merely as a background trend that affected all lines of living development. Instead of being the logical goal of the whole process, man became merely the highest point so far reached by any of these lines. Darwin's theory heavily reinforced this trend and disposed of the whole idea of a preordained plan of growth. But, encouraged by progressionist philosophies such as Spencer's, the theory allowed a reinterpretation that fitted in well with the late nineteenth century's passionate faith in the naturally progressive character of the universe in general and man in particular. The present century's lack of interest in progression can in part be attributed to the ever-increasing difficulty of defining the concept against the background of a complex and irregular fossil record. But we may suspect that the collapse of the idea

First I've heard of it!

of biological progression would not have occurred had not the modern period become generally disillusioned with the naively optimistic philosophies of social progress that were so common in the previous century.

Notes

[1]T.H. Huxley's rejection of progression is discussed below, along with a brief reference to Julian Huxley's contributions to modern evolutionary thought.

[2]See Charles Darwin, *Journal of Researches into the Natural History and Geology of the countries visited on the voyage of H.M.S. Beagle around the world*, 2nd edition, p. 125. For a discussion of Darwin's paleontological work, see A.F. Romer, "Darwin and the fossil record."

[3]Owen's claim is discussed by Darwin in a letter to Lyell in 1859; see Francis Darwin (ed.), *More Letters of Charles Darwin. A record of his work in a series of hitherto unpublished letters*, I, p. 133.

[4]See the letter to J.D. Hooker, July 22nd, 1879, *More Letters*, II, pp. 20–22. Darwin was also interested in a suggestion by Gaston de Saporta that the evolution of the mammals had been delayed until the appearance of the appropriate kinds of vegetation, although this did not involve a directional theory of the earth; see his letter to Saporta of December 24th, 1877, *Life and Letters of Darwin*, III, pp. 284–286. See also Conry, *Correspondence entre Charles Darwin et Gaston de Saporta*.

[5]Darwin to Hooker, 1854, *More Letters*, I, p. 76.

[6]Charles Darwin, "Essay of 1844," in Darwin and Wallace, *Evolution by Natural Selection* with a foreword by Sir Gavin de Beer, p. 157.

[7]See the note in *More Letters*, I, p. 115. Darwin used the sixth (1847) edition of *Vestiges*; on his general reaction to the work, see Egerton, "Refutation and conjecture: Darwin's response to Sedgwick's attack on Chambers."

[8]See the letter to Lyell, March 12th, 1873, *Life and Letters of Darwin*, II, p. 14.

[9]Darwin, "Essay of 1844," 231.

[10]Darwin to Hooker, December 30th, 1858, *More Letters*, I, p. 115.

[11]*Ibid.*

[12]Charles Darwin, *On the Origin of Species by means of Natural Selection, or the preservation of favoured races in the struggle for life*, p. 345.

[13]*Ibid.*, p. 489.

[14]See Maurice Mandelbaum, *History, Man, and Reason, a study in nineteenth century thought*, pp. 77–92.

[15]See Jane Oppenheimer, "An embryological enigma in the *Origin of Species*." Darwin refers to Agassiz's ideas on recapitulation in the *Origin*, expressing the hope that they will be confirmed; see 338 and 450.

[16]See the English translation of Ernst Haeckel's *Natürliche Schöpfungs-Geschichte: The History of Creation; or the development of the earth and its inhabitants by the action of*

natural causes, I, pp. 309–312. Haeckel was well aware of the branching nature of evolution, but was firmly convinced that the fossil record showed the existence of a progressive trend such as that postulated by Bronn; see *ibid.*, I, p. 277. E.S. Russell points out that in theory the biogenetic law was not equivalent to the old recapitulation theory, although it soon became identical in its applications; see Russell, *Form and Function*, pp. 252–257.

[17] Darwin, "Essay of 1844," 223–230 and *Origin*, pp. 439–440.

[18] Darwin, *Origin of Species*, 4th edition (1866), p. 531. The corresponding passage from the "Essay of 1844," p. 233, quoted by Oppenheimer, states more clearly that the original vertebrate must have been a fish, but still does not support the law of parallelism. Müller's *Für Darwin* was translated into English as *Facts and Arguments for Darwin* (1869); for a description of his work see Russell, *Form and Function*, pp. 252–253. Darwin's discussion of Müller's work in the fourth edition has led Sir Gavin de Beer to follow Oppenheimer in claiming that Darwin had moved away from von Baer's position; see his "Darwin and embryology," p. 157. Following Müller, Darwin argued that the modern Nauplius must resemble the ancestral form of a number of crustacean groups, each of which passes through a "Nauplius stage" in the course of its growth. De Beer points out that Nauplius is a highly specialized form—but Darwin makes it clear that *he* believed it to be a more generalized type, and thus his speculations are again not incompatible with von Baer's concept of development.

[19] See Russell, *Form and Function*, p. 237.

[20] For references to Carpenter's work, see above, Chapter 5, note 30. It is strange that Oppenheimer discusses Carpenter's account of von Baer's embryology at some length, without mentioning that he held similar patterns of development to be observable in the fossil record.

[21] See Agassiz, *Essay on Classification*, pp. 246–252, for an exposition of von Baer's system.

[22] Darwin, *Origin of Species* (1859), pp. 305–306.

[23] *Ibid.*, pp. 306–310. It is significant that Darwin did not mention the appearance of the first fish in the Silurian as a particular source of difficulty—a sign that discontinuous progression was no longer a force to be reckoned with.

[24] See W.E. Logan, "On the occurrence of organic remains in the Laurentian rock of Upper Canada." This paper is followed by supporting accounts written by Dawson and Carpenter. On the subsequent debate over the organic nature of *Eozoön*, see Charles F. O'Brien, "*Eozoön canadense*, the dawn animal of Canada." The Foraminifera are a group of protozoa with a shell, but no obvious internal structure.

[25] Darwin, *Origin of Species* (1866), p. 371.

[26] See the note in *Life and Letters of Darwin*, I, p. 333.

[27] Darwin, *Origin of Species* (1859), pp. 329–330.

[28] *Ibid.*, p. 313.

[29] See Darwin, *Origin of Species* (1866), p. 402. Darwin was referring to Carpenter's *Introduction to the Study of the Foraminifera*. When the *Athenaeum* review of this work accused Carpenter of being a Darwinian, he retorted with a letter emphasizing that his researches showed the Foraminifera never to have evolved into anything more advanced;

see the *Athenaeum* No. 1848 (March 28th, 1863), 417–419 and No. 1849 (April 4th), 461. The story of this incident is given in *Life and Letters of Darwin*, III, pp. 17–18.

[30] Darwin, *Origin of Species* (1859), pp. 334–335. Although his work had been used by Sedgwick against *Vestiges*, Falconer now admitted that persistence of type among the elephants was not a valid argument against Darwinism; see his "On the American fossil elephants of the region bordering on the Gulph of Mexico . . .," especially 83.

[31] Darwin, *Origin of Species* (1859), p. 341.

[32] See H.L. McKinney, *Wallace and Natural Selection*, p. 41.

[33] A.R. Wallace, "On the law which has regulated the introduction of new species," see 11–13.

[34] Wallace, "On the tendency of varieties to depart indefinitely from the original type," see the section "Superior varieties will ultimately extirpate the original species."

[35] *Ibid.*, p. 33.

[36] Wallace, "Sir Charles Lyell on geological climates and the origin of species," see 203.

[37] See for instance his *Darwinism, an exposition of the theory of Natural Selection with some of its applications*, pp. 375–409.

[38] This position is developed in the review of Lyell cited above and in Wallace's "The limits of natural selection as applied to man."

[39] See Alvar Ellegård, *Darwin and the General Reader. The reception of Darwin's theory of evolution in the British periodical press. 1859–1872*, pp. 272–273.

[40] There is no general and sympathetic study of this last phase of theistic evolutionism, although there is an account of Mivart's work by Jacob Gruber, *A Conscience in Conflict. The life of St. George Mivart.* For the relevant works, see the essays in W.B. Carpenter's *Nature and Man;* Mivart's *Genesis of Species* (London, 1871) and the Duke of Argyll's *The Reign of Law* (London, 1867).

[41] See Spencer's *Principles of Biology* (1864, originally issued 1863–64), I, pp. 420–421, and more especially the article "Illogical geology" (1859).

[42] See Spencer's "The development hypothesis." He mentions his distrust of *Vestiges'* direct progressionism in the autobiographical "The filiation of ideas," in David Duncan (ed.), *The Life and Letters of Herbert Spencer*, p. 541. As early as 1844 he had suggested the progress of life might be caused by changing physical conditions; see the article reprinted in his *Autobiography*, I, pp. 533–537.

[43] See Spencer, *First Principles of a New Philosophy* (1864, issued 1860–62), p. 404. The same point had already been made in the essay "Progress: its law and cause" (1857), see p. 48.

[44] On Spencer's support for the inheritance of acquired characteristics, see his *Principles of Biology*, I, pp. 402–463. For a recent analysis of his social philosophy, see J.D.Y. Peel, *Herbert Spencer, the evolution of a sociologist.*

[45] See my "The changing meaning of 'evolution.'"

[46] See R.H. Lock, *Recent Progress in the Study of Variety, Heredity and Evolution*, pp. 21–22. By this time, however, the general popularity of Spencer's philosophy in England had receded.

[47]Alexander Agassiz, "Paleontological and embryological development," see 388. On Agassiz's life, see G.R. Agassiz, *Letters and Recollections of Alexander Agassiz, with a sketch of his life and work*. E.S. Russell suggests that there was indeed a slackening of interest in the reconstruction of the history of life toward the end of the century; see *Form and Function*, p. 268 and p. 302.

[48]Darwin to Alexander Agassiz, May 5th, 1881, *Life and Letters of Darwin*, III, p. 245.

[49]E.D. Cope, "On the origin of genera," 243–244 and 269. This article is reprinted in Cope, *The Origin of the Fittest. Essays in evolution*, pp. 41–123.

[50]On Cope's Lamarckian theory, see his "The laws of organic development," (1871), and "The method of creation of organic forms," (1873), reprinted in *Origin of the Fittest*, pp. 173–214.

[51]For a discussion of Hyatt's views, see the biography by Stephen Jay Gould in the *Dictionary of Scientific Biography*. Darwin corresponded extensively with Hyatt on the law of acceleration and on Cope's use of it; see *More Letters of Darwin*, I, pp. 338–348. Elsewhere he complained that he despaired of ever understanding their meaning; see *Life and Letters of Darwin*, III, p. 233.

[52]See Ernst Haeckel, *History of Creation*, II, p. 248. This point was used against Darwin by St. George Jackson Mivart in his *The Genesis of Species*, pp. 76–81. Mivart and Haeckel both mention T. H. Huxley as a supporter of this interpretation of the origin of the mammals, Mivart citing his (unpublished) Hunterian Lectures for 1866. As yet I have not noticed the point developed in Huxley's published essays and papers, although he did advance other important ideas on the origin of the mammals which are mentioned below. For a modern comment on the belief that the mammals are "polyphyletic," see George Gaylord Simpson, "Mesozoic mammals revisited," 192–193.

[53]For a summary of Huxley's early views on the evolution of the horse, see his 1870 presidential address to the Geological Society, reprinted under the title "Palaeontology and evolution," especially 355–360.

[54]On Huxley's visit to Marsh, see Leonard Huxley (ed.), *Life and Letters of Thomas Henry Huxley*, II, pp. 202–205. Marsh's discoveries are outlined in his "Notice of some new equine mammals from the Tertiary formation." On Marsh's life, see Charles Schuchert and Clara May LeVene, *O.C. Marsh, pioneer in paleontology*.

[55]Huxley, *American Addresses, with a lecture on the study of biology*, pp. 85–90 (first edition published in 1877).

[56]See *ibid.*, p. 90; and Marsh, "Notice of some new Tertiary mammals."

[57]See Marsh, *Dinocerata. A monograph of an extinct order of gigantic mammals*, p. 173. For Huxley's speculations along similar lines, see his "Palaeontology and evolution," pp. 362–364.

[58]See Russell, *Form and Function*, pp. 357–364.

[59]See T.H. Huxley, "On the characters of the pelvis in the mammalia, and the conclusions respecting the origin of the mammals which may be based on them."

[60]Thus in 1888 H.G. Seeley identified *Pareiasaurus* as a form intermediate between the amphibians and reptiles, yet already possessing some mammalian characteristics; he suggested that it was a possible starting point from which the Sauropsida and the

mammals had diverged. See "Researches on the structure, organization, and classification of the fossil Reptilia.—II, on *Pareiasaurus bombidens* (Owen), and the significance of its affinities to amphibians, reptiles and mammals," especially 106–109.

[61] See, for instance, Seeley, "Researches . . .—III, On parts of the skeleton of a mammal from the Triassic rocks of Klipfontein, Fraserberg, S. Africa (*Theriodesmus phylarchus*, Seeley) illustrating the reptilian inheritance of the mammalian hand." This particular form was later shown to be a reptile with mammalian characteristics; see "Researches . . .—IX section 2, The reputed mammals from the Karoo formation of the Cape Colony." On the background to Seeley's work, see W.E. Swinton, "Harry Govier Seeley and the Karoo reptiles."

[62] T.H. Huxley, "On the animals which are most nearly intermediate between birds and reptiles," see 311.

[63] See H.G. Seeley, "On the classification of the fossil animals commonly named Dinosauria."

[64] Richard Owen, "On the *Archaeopterix* of von Meyer. . . ." Von Meyer had described a feather from the form discovered in 1862.

[65] Huxley, "Remarks on *Archaeopterix lithographica*." In "The animals most nearly intermediate between birds and reptiles," Huxley did at least admit that there were reptilian features in the tail, although in general he repeated his claim that *Archaeopterix* was just a bird; see p. 307.

[66] See the section on the evidences of evolution in Huxley's *American Addresses*.

[67] Marsh, "Preliminary description of *Hesperornis regalis*. . . ." See also "On a new subclass of fossil birds (Odontornithes);" "On the Odontornithes, or birds with teeth;" and the monograph *Odontornithes; a monograph on the extinct toothed birds of North America.*

[68] Marsh, *Odontornithes*, p. 188.

[69] *Ibid.*, p. 185.

[70] Marsh, "Recent discoveries of extinct mammals."

[71] Marsh, *Dinocerata*, p. 59.

[72] *Ibid.*

[73] Marsh, *Introduction and Succession of Vertebrate Life in America. An address delivered before the American Association for the Advancement of Science at Nashville, Tenn., August 30, 1877*, p. 55. No place or date of publication for this monograph is given; the address was also published serially in *Nature* for 1877.

[74] Haeckel had actually suggested the name *Pithecanthropus* for the hypothetical man-ape form proposed in works such as his *Anthropogenie, oder Entwickelungsgeschichte des Menschens,* translated as *The Evolution of Man.* Java man, *Pithecanthropus erectus*, was discovered by Eugene Dubois in 1894; Haeckel was a leading figure urging its acceptance by the scientific world; see his *The Last Link. Our present knowledge of the descent of man.*

[75] T.H. Huxley, *Evidences as to Man's Place in Nature*, p. 108.

[76] Huxley, "Geological contemporaneity and persistent types of life," see p. 286.

[77]*Ibid.*, pp. 290–291. For Hooker's own comments on the lack of fossil evidence for evolutionary progression, see the introduction to his *Flora of Tasmania*; I have used the version reprinted in the *American Journal of Science*: "On the origination and distribution of vegetable species:—Introductory Essay to the Flora of Tasmania," see 308–313.

[78]Huxley, "Geological contemporaneity and persistent types of life," 292.

[79]*Ibid.*, 303.

[80]Huxley, "Palaeontology and evolution," 347–366.

[81]On orthogenesis see Julian Huxley, *Evolution: the modern synthesis*, pp. 505–516.

[82]See Julian Huxley, *Evolution in Action, based on the Patten Foundation lectures delivered at the University of Indiana in 1951*, pp. 11–12 and *Evolution: the modern synthesis*, pp. 556–578; G.G. Simpson, *The Meaning of Evolution. A study in the history of life and its significance for man*, pp. 261–262; J.M. Thoday, "Natural selection and biological progress." For a survey of these developments, see T.A. Gouge, *The Ascent of Life. A philosophical study of the theory of evolution*, Chapter 5.

[83]See Pierre Teilhard de Chardin, *The Phenomenon of Man*. Julian Huxley expressed support for Teilhard's ideas in his introduction to this English translation, but other scientists have been suspicious or violently hostile; see for instance the critiques by G. G. Simpson and Peter Medawar, reprinted along with Huxley's comments and a selection from Teilhard's work in Philip Appleman (ed.), *Darwin, a Norton critical edition*, pp. 458–485.

Plates

PLATE VIII

Skeleton of *Apateon* (von Meyer) or *Archegosaurus* (Goldfuss), from the upper Carboniferous rocks of Bavaria. This reconstruction is taken from Richard Owen's *Palaeontology* (1860), p. 169. After some debate Owen concluded that this was a "reptile" allied to the perrenibranchiate amphibians, and here he gives the structure of the modern *Proteus* for comparison.

PLATE IX

Diagram showing the geological distribution of the various orders of reptiles, from Owen's *Palaeontology*, p. 285.

PLATE X

The "tree of life" from H. G. Bronn's *Essai*, p. 524 (*Comptes rendus*, 907).

PLATE XI

Part of the diagram used by Darwin in the *Origin of Species* to illustrate the principle of divergence. Some forms inhabit stable environments and do not change (vertical lines), but when subjected to changing conditions a species tends to form numerous varieties. Often only one of these survives, but sometimes the two most extreme varieties are perpetuated to form two new species.

PLATE XII

Darwin's views of the two possible ways in which the mammals could have developed. In each case "A" represents a hypothetical ancestral form. From Darwin's letter to Lyell, September 23rd, 1860, *Life and Letters of Darwin*, II, p. 343.

PLATE XIII

Modification of the teeth and lower parts of the limbs from the four-toed *Orohippus* of the Eocene to the modern horse. This sequence as reconstructed by Marsh and Huxley was very popular among evolutionists, and the diagram was often reproduced. This example is taken from A. R. Wallace's *Darwinism* (1889), p. 388.

PLATE XIV

Skeleton of *Hesperornis regalis,* a toothed bird from the Cretaceous rocks of Kansas, as reconstructed by Marsh. From Huxley's *American Addresses* (1888), p. 52.

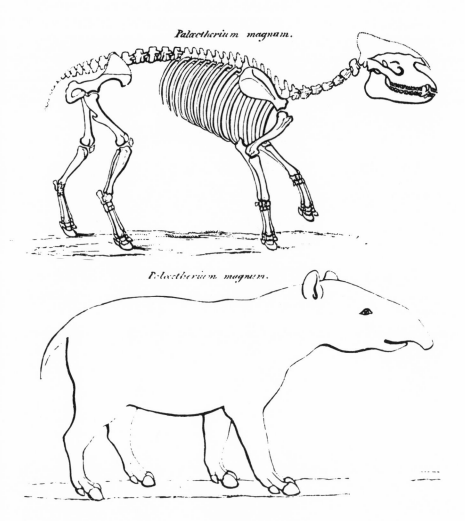

Palaeotherium magnum.

Palaeotherium magnum.

PLATE I

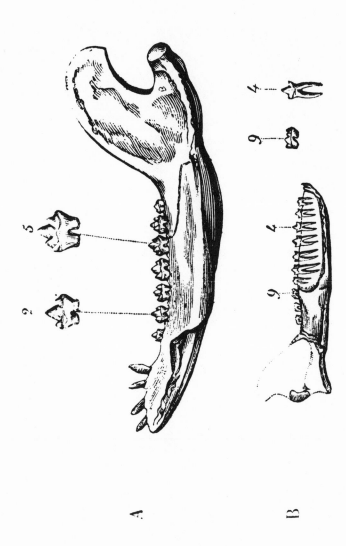

Lower Jaws and Teeth of Didelphys from the Oolite of Stonesfield, Oxon.

PLATE II

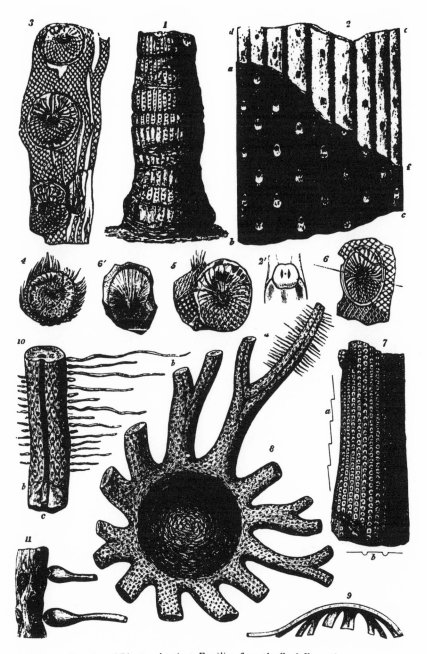

Remains of Plants, of extinct Families, from the Coal Formation.

PLATE III

Pl. 16.

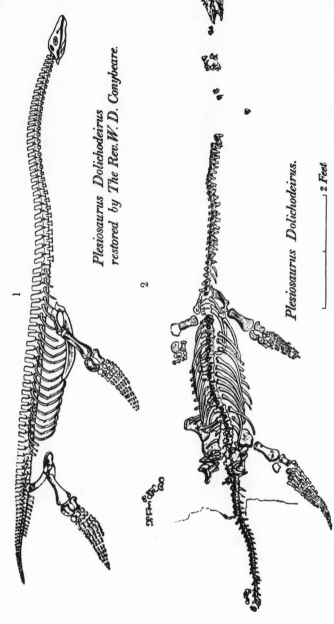

1

Plesiosaurus Dolichodeirus
restored by The Rev.W.D.Conybeare.

2

Plesiosaurus Dolichodeirus.

|_____| 2 *Feet*

PLATE IV

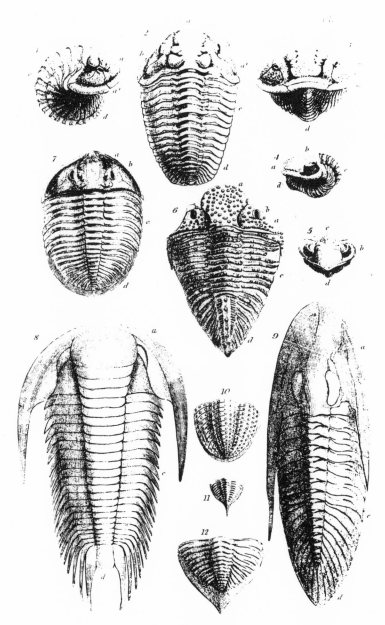

TRILOBITES.

PLATE V

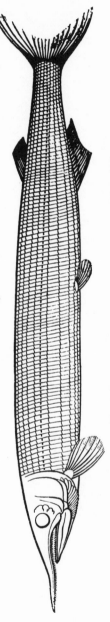

1

c

Lepidosteus osseus.
living in rivers of N. America.

2
NS

b a

3

Lepidosteus osseus, lower Jaw : nat. size.
from a young animal.

4

d

c

b

a

Megalichthys Hibberti, upper Jaw : nat. size.
from a young animal.

5

Fossil Aspidorhynchus, from the Jura limestone of Solenhofen.

Examples of recent and fossil Sauroid Fishes.

PLATE VI

PLATE X.

RESTORATION OF CEPHALASPIS.

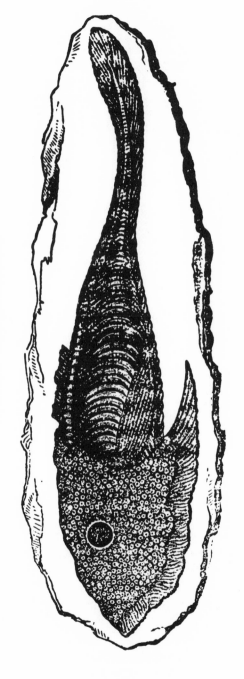

PLATE VII

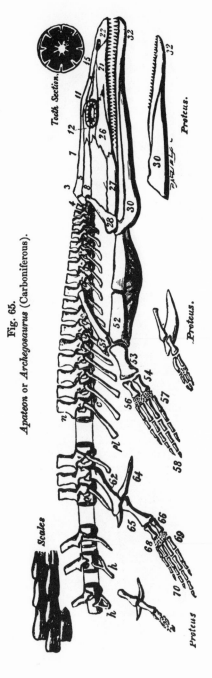

Fig. 65.

Apateon or Archegosaurus (Carboniferous).

Teeth Section.

Proteus.

Proteus.

Scales

Proteus

PLATE VIII

Table of Geological Distribution of Reptilia.

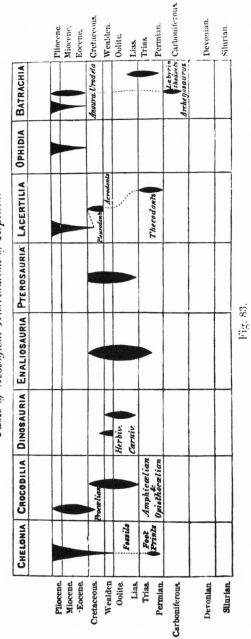

Fig. 83.

PLATE IX

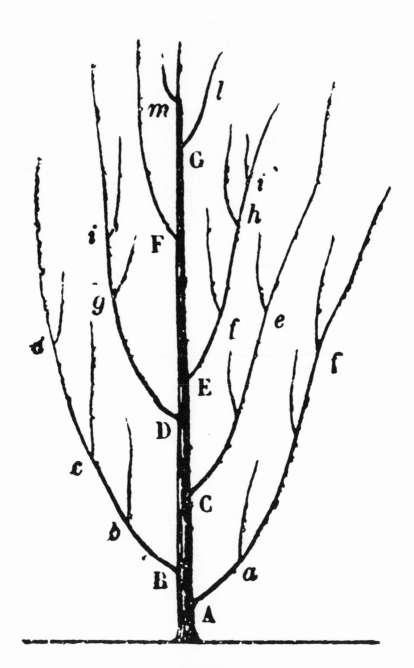

PLATE X

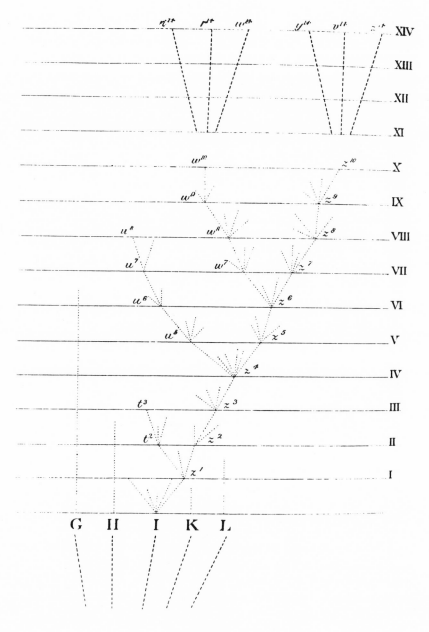

PLATE XI

DIAGRAM I.

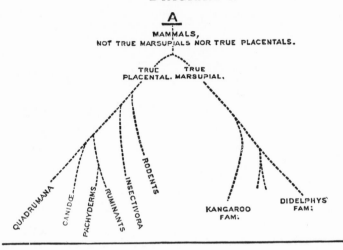

A

MAMMALS,
NOT TRUE MARSUPIALS NOR TRUE PLACENTALS.

TRUE
PLACENTAL.

TRUE
MARSUPIAL.

RODENTS

QUADRUMANA

CANIDÆ

PACHYDERMS

RUMINANTS

INSECTIVORA

KANGAROO
FAM:

DIDELPHYS
FAM:

DIAGRAM II.

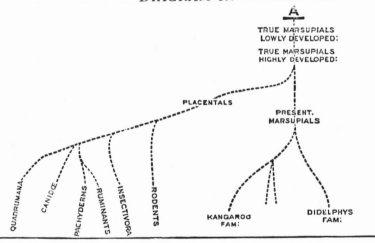

A

TRUE MARSUPIALS
LOWLY DEVELOPED:

TRUE MARSUPIALS
HIGHLY DEVELOPED:

PLACENTALS

PRESENT.
MARSUPIALS

QUADRUMANA

CANIDÆ

PACHYDERMS

RUMINANTS

INSECTIVORA

RODENTS

KANGAROO
FAM:

DIDELPHYS
FAM:

PLATE XII

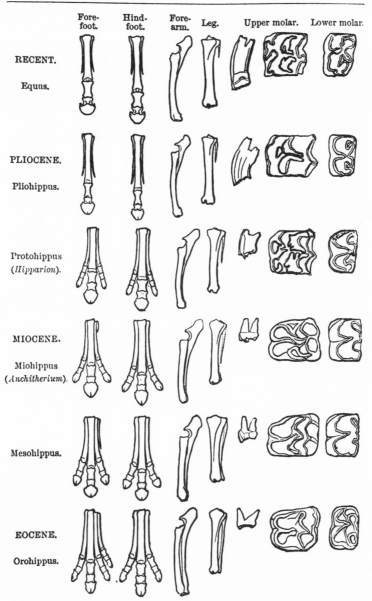

Fig. 33.—Geological development of the horse tribe (Eohippus since discovered).

PLATE XIII

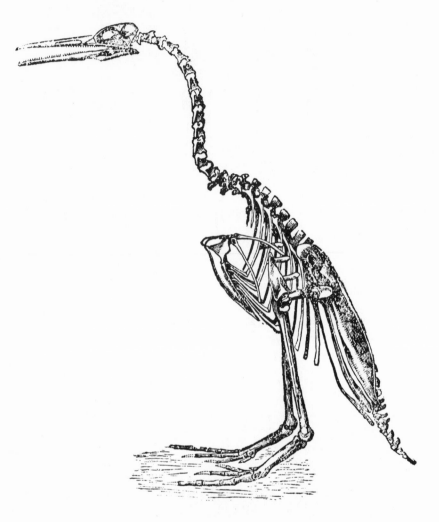

PLATE XIV

Bibliography

This bibliography is simply a list of the books and articles used in the preparation of this study; it does not pretend to indicate all of the paleontological material that is indirectly related to progressionism. Some of the more important works relevant to eighteenth century evolutionism have been included.

Primary Sources

Agassiz, Alexander. "Paleontological and embryological development."*Am. J. Sci.* XX (whole number CXX), 1880: 294–302 and 375–389.

Agassiz, Elizabeth Cary. *Louis Agassiz, his life and correspondence.* 2 vols. London, 1885.

Agassiz, G.R. *Letters and Recollections of Alexander Agassiz, with a sketch of his life and work.* Boston, 1913.

Agassiz, Louis. *Recherches sur les poissons fossiles . . .* 5 vols. of text, plus plates. Neuchatel, 1833–43.

———. "On a new classification of fishes and on the geological distribution of fossil fishes." *Edinb. New Phil. J.* XVIII (135): 175–178.

———. Note on De Blainville's "Doutes sur le pretendue Didelphe fossile de Stonesfield," *Comptes rendus* VII (1838): 537.

———. *Monographie d'échinoderms, vivans et fossiles.* Neuchatel, 1841

———. "On the succession and development of organized beings at the surface of the terrestrial globe; being a discourse delivered at the inauguration of the Academy of Neuchatel," *Edinb. New Phil. J.* XXXIII (1842): 388–399.

———. "Report on the fossil fish of the Devonian system or Old Red Sandstone." In *Report . . . of the British Association for the Advancement of Science . . . 1842*, pp. 80–88, London, 1843.

———. *Monographie des poissons fossiles du vieux grès rouge, ou systeme devonien . . .* Neuchatel, 1844.

———. *Twelve Lectures on Comparative Embryology, delivered before the Lowell Institute in Boston, December and January, 1848–49.* Boston, 1849.

———. "On the difference between progressive, embryonic and prophetic types in the succession of organized beings through the whole range of geological time." *In Proceedings of the American Association for the Advancement of Science . . . 1849*, Boston, 1850; reprinted 1855, pp. 432–438.

165

————. *Essay on Classification.* Edited by Edward Lurie. Originally Vol. I of *Contributions to the Natural History of the United States, 1857.* Cambridge, Mass.: Harvard University Press, 1962.

————. "Professor Agassiz on the origin of species." *Am. J. Sci.* 2nd series, XXX (1860): 142–154.

————. *The Structure of Animal Life. Six Lectures delivered at the Brooklyn Academy of Music in January and February, 1862,* New York, 1866.

————. "Evolution and permanence of type." *Atlantic Monthly* XXXIII (1874): 92–101.

————. *Studies on Glaciers.* Translated by Albert V. Carozzi. New York: Hafner, 1967.

Agassiz, Louis, and Gould, A.A. *Principles of Zoology, touching the structure, development, distribution and natural arrangement of animals, living and extinct.* Revised edition. Boston, 1861.

Anon. "Geology versus development." *Fraser's Magazine* XLII (1850): 355–371.

Ansted, David Thomas. *The Ancient World; or, picturesque sketches of creation.* London, 1847.

Argyll, G.D. Campbell, Duke of. *The Reign of Law.* London, 1867.

————. *The Unity of Nature.* 3rd edition. London, 1888.

Babbage, Charles. *The Ninth Bridgewater Treatise, a fragment.* 2nd edition. London, 1838.

Baer, Karl Ernst von. *Über Entwickelungsgeschichte der Thiere. Beobachtung und Reflexion.* Erster theil, Konigsberg, 1828. Reprinted Brussels: Culture et Civilization, 1967.

Bayne, Peter. *The Life and Letters of Hugh Miller.* 2 vols. London, 1871.

Bonnet, Charles. *Oeuvres d'histoire naturelle et de philosophie.* 19 vols. Neuchatel, 1779.

Brongniart, Adolphe. *Histoire des végétaux fossiles, ou recherches botaniques et géologiques sur les végétaux renfermés dans les diverses couches du globe.* 2 vols. Paris and Amsterdam, 1828–37.

————. *Prodrome d'une histoire des végétaux fossiles* . . . Paris and Strasburg, 1828

————. "Considérations générales sur la nature de la végétation qui couvrait la surface de la terre aux diverses époques de la formation de son écorce." *Ann. des sci. nat.* XV (1828): 225–258. Printed separately as a pamphlet, Paris, 1828.

————. "General considerations on the nature of the vegetation which covered the surface of the earth at the different epochs of the formation of its crust . . ." *Edinb. New Phil. J.* VI (1829): 349–371.

Bronn, Heinrich Georg. "Some considerations on palaeontological statistics, drawn up from the *History of Nature (Geschichte der Natur)* or *Index Palaeontologicus* by Prof. H.G. Bronn." *Quart. J. Geol. Soc. Lond.* V (1849) part 2: 38–58.

―――. *Essai d'une réponse a la question de prix proposée en 1850 par l'Académie des Sciences* . . . , *savoir:—Etudier les lois de la distribution des corps organisés fossiles* . . . Paris, 1861. Also printed *Comptes rendus* supp. vol. II (1861): 377–918.

―――. "On the laws of evolution of the organic world during the formation of the crust of the earth." *Ann. and Mag. Nat. Hist.* 3rd series, IV (1859): 81–90 and 175–184.

―――. "Notice of *Untersuchungen* . . ." *Quart. J. Geol. Soc. Lond.* XV (1859) part 2: 1–5.

Buckland, William. *Vindiciae Geologicae, or the connexion of geology with religion explained.* Oxford, 1820.

―――. "An account of an assemblage of fossil teeth and bones . . . discovered in a cave at Kirkdale, Yorkshire, in 1821." *Phil. Trans.* CXII (1822): 171–235.

―――. *Reliquiae Diluvianae, or observations on the organic remains contained in caves, fissures and diluvial gravel, and other geological phenomena, attesting the action of a universal deluge.* 2nd edition. London, 1824.

―――. "Notice on the *Megalosaurus*, or great fossil lizard of Stonesfield." *Trans. Geol. Soc. Lond.* 2nd series, I (1824): 390–396.

―――. *Geology and Mineralogy considered with reference to Natural Theology.* 2 vols. 2nd edition. London, 1837.

Buffon, Georges Louis Leclerc, Comte de. *Histoire naturelle, générale et particuliere.* 15 vols. Paris, 1749–67.

―――. *Histoire naturelle des oiseaux.* 9 vols. Paris, 1770–83.

―――. *Histoire naturelle, générale et particulière, Supplément.* 7 vols. Paris, 1774–84.

―――. *Les Epoques de la nature.* Edited by J. Roger. *Mémoires du Muséum National d'Histoire naturelle*, nouvelle série, série C, X. Paris, 1962.

Burnet, Thomas. *The Sacred Theory of the Earth.* London, 1691. Reprinted with an introduction by Basil Willey. London: Centaur Press, 1965.

Burton, Charles. *Lectures on the World before the Flood.* London, 1844.

Carpenter, William Benjamin. *Principles of Physiology, General and Comparative.* 3rd edition. London, 1851.

―――. *Principles of Comparative Physiology.* American edition from the 4th London edition. Philadelphia. 1854.

———. "Researches on Foraminifera." *Phil. Trans.* CXLVI (1856): 181–236.

———. "The theory of development in nature." *Brit. and For. Med. Chirurg. Rev.* XXV (1860): 269–295.

———. "On the Origin of Species." *National Review* X (1860): 188–214.

———. *Introduction to the Study of the Foraminifera.* London, 1862.

———. "On the structure and affinities of *Eozoön canadense.*" *Proc. Roy. Soc.* XIII (1864): 545–549.

———. "Additional notes on the structure and affinities of *Eozoön canadense.*" *Quart. J. Geol. Soc. Lond.* XXI (1865): 59–66.

———. "Supplementary notes on the structure and affiliation of *Eozoön canadense.*" *Quart. J. Geol. Soc. Lond.* XXII (1866): 219–228.

———. Presidential address. In *Report . . . of the British Association for the Advancement of Science . . . 1872.* Pp. lxix–lxxiv. London, 1873.

———. *Nature and Man. Essays scientific and philosophical.* With an introductory memoir by J. Erstlin Carpenter. New York, 1889.

Chambers, Robert. *Vestiges of the Natural History of Creation.* London, 1844. Reprinted with an introduction by Sir Gavin de Beer. Leicester: University Press, 1969.

———. *Vestiges of the Natural History of Creation.* 4th edition. London, 1845. 12th edition with an introduction by Alexander Ireland. Edinburgh, 1884.

———. *Explanations: a sequel to the Vestiges of the Natural History of Creation.* 2nd edition. London, 1846.

Clark, J.W., and Hughes, T.M. *The Life and Letters of the Rev. Adam Sedgwick.* 2 vols. Cambridge, 1890.

Conybeare, W.D. "Additional notes on the fossil genera *Ichthyosaurus* and *Plesiosaurus.*" *Trans. Geol. Soc. Lond.* 2nd series, I (1824): 103–123.

———. "On the discovery of an almost perfect skeleton of the *Plesiosaurus.*" *Trans. Geol. Soc. Lond.* 2nd series, I (1824): 381–389. Reprinted *Phil. Mag.* LXV (1825): 412–420.

———. "Letter on Lyell's *Principles of Geology.*" *Phil. Mag.* 2nd series, VIII (1830): 215–219.

———. "An examination of those phaenomena of geology which seem to bear most directly on theoretical speculation." *Phil. Mag.* 2nd series, VIII (1830): 359–362 and IX (1831): 19–23, 111–117, 188–197, and 258–270.

———. "Report on the progress, actual state and ulterior prospects of geological science." In *Report . . . of the British Association for the Advancement of Science . . . 1831 and 1832.* Pp. 365–414. 2nd edition, London, 1835.

———. (M.J.S. Rudwick). "A critique of uniformitarian geology: a letter from W.D. Conybeare to Charles Lyell, 1841." *Proc. Am. Phil. Soc.* CXI (1967): 272–281.

Conybeare, W.D., and De La Beche, H.T. "Notice of a new fossil animal forming a link between the *Ichthyosaurus* and the crocodile." *Trans. Geol. Soc. Lond.* V (1821): 559–594.

Conybeare, W.D., and Phillips, William. *Outline of the Geology of England and Wales.* London, 1822.

Cope, E.D. "On the origin of genera." *Proc. Acad. Nat. Sci., Philadelphia* XX (1868): 242–300.

———. "The laws of organic development." *American Naturalist* V (1871): 593–605.

———. "The method of creation of organic forms." *Proc. Am. Phil. Soc.* XII (1873): 229–265.

———. *The Origin of the Fittest. Essays in evolution.* New York, 1887.

Cuvier, Georges, baron. *Recherches sur les ossemens fossiles de quadrupèdes, ou l'on retablit les characteres de plusieurs espèces d'animaux que les révolutions du globe paroissent avoir disparu.* 4 vols. Paris, 1812. 4th edition. 12 vols. Paris, 1836.

———. "Sur un nouveau rapprochement à établir entre les classes qui composent le règne animal." *Ann. Mus. Hist. Nat., Paris* XII (1812): 73–84.

———. *Essay on the Theory of the Earth, translated from the French . . . by Robert Kerr, with mineralogical notes by Professor Jameson.* Edinburgh, 1813. 5th edition. London, 1825.

———. *Le Regne animal distribué d'après son organisation, pour servir de base a l'histoire naturelle des animaux et d'introduction à l'anatomie comparée.* 4 vols. Paris, 1817.

Cuvier, Georges, and Brongniart, Alexandre. *Essai sur la géographie minéralogique des environs de Paris, avec une carte géognostique, et des coupes de terrain.* Paris, 1811.

Dana, J. D. "Agassiz's Contributions to the Natural History of the United States." *Am. J. Sci.* 2nd series, XXV (1858): 202–216 and 321–341.

———. *A Manual of Geology.* Philadelphia and London, 1862.

Darwin, Charles Robert. *Journal of Researches into the Natural History and Geology of the countries visited during the voyage of H.M.S. Beagle around the world . . .* London, 1845. Reprinted 1891.

———. *On the Origin of Species by means of Natural Selection, or the preservation of favoured races in the struggle for life.* London, 1859. 4th edition. London, 1866.

———. *The Descent of Man and selection in relation to sex.* 2nd edition revised. London, 1885.

———. (Conry, Y.). *Correspondence entre Charles Darwin et Gaston de Saporta, précedé d'une histoire de la paléobotanique en France en XIXe siecle.* Paris: P.U.F., 1972.

Darwin, Charles, and Wallace, Alfred Russell. *Evolution by Natural Selection.* With a foreword by Sir Gavin de Beer. Cambridge: University Press, 1958.

Darwin, Erasmus. *Zoonomia, or the laws of organic life.* 2 vols. London, 1794.

Darwin, Francis. *The Life and Letters of Charles Darwin.* 3 vols. London, 1887.

———. *More Letters of Charles Darwin, a record of his work in a series of hitherto unpublished letters.* 2 vols. New York, 1903.

Davy, Sir Humphrey. *Consolations in Travel, or the last days of a philosopher.* 5th edition. London, 1851.

Dawson, J.W. "On a terrestrial mollusk, a chilognathus Myriapod and some new species of reptiles from the coal formations of Nova Scotia." *Quart. J. Geol. Soc. Lond.* XVI (1860): 268–277.

———. "Review of Darwin on the Origin of Species by means of natural selection." *Canadian Naturalist and Geologist* V (1860): 100–120.

———. "On the structure of certain organic remains in the Laurentian limestone of Canada." *Quart. J. Geol. Soc. Lond.* XXI (1865): 51–59.

———. "Modern ideas of derivation." *Canadian Naturalist.* New series, IV (1869): 121–138.

———. "On the bearing of Devonian botany on questions as to the origin and extinction of species." *Am. J. Sci.* 3rd series, II, whole no. CII (1871): 410–416.

———. *Life's Dawn on Earth, being the history of the oldest known fossil remains and their relations to geological time and the development of the animal kingdom.* London, 1875. Also printed under the title *The Dawn of Life.* Montreal, 1875.

———. *Some Salient Points in the Science of the Earth.* Montreal, 1893.

De Blainville, H.D. "Doutes sur le pretendue Didelphe fossile de Stonesfield . . ." *Comptes rendus* VII (1838): 402–448.

———. "Nouveau doutes . . ." *Comptes rendus* VII (1838): 727–736.

De La Beche, Henry T. *A Geological Manual.* 3rd edition. London, 1833.

Deluc, Jean André. "Letters to Dr. James Hutton, F.R.S. Ed., on his theory of the earth." *Monthly Review* 2nd series, II (1790): 206–227 and 582–601; III (1790): 573–586; IV (1791): 564–585.

———. *Lettres sur l'histoire physique de la terre, addressées a M. le Professeur Blumenbach.* Paris, 1798.

Diderot, Denis. *Oeuvres complètes.* Edited by J. Assezat. 20 vols. Paris, 1875–77.

D'Orbigny, Alcide. *Paléontologie francaise. Description zoologique et géologique de tous les animaux mollusques et rayonnés fossiles de France. Terrains crétacé.* 6 vols. (vols. 7 and 8 completed by other authors.) Paris, 1840–47. *Terrains oolitiques ou jurassiques.* (later vols. added by other authors.) Paris, 1842.

———. *Prodrome de paléontologie stratigraphique universelle des animaux mollusques et rayonnés, faisant suite au cours elementaire de paléontologie et de géologie stratigraphique.* 3 vols. Paris, 1849–52.

———. "Recherches zoologiques sur l'instant d'apparition, dans les ages du monde, des ordres d'animaux, comparé au degré de perfection de l'ensemble de leurs organes." *Ann. des sci. nat.* 3rd series, XIII (1850): 228–236.

———. "Recherches zoologiques sur la marche successive de l'animalisation à la surface du globe, depuis les temps zoologiques les plus anciens, jusqu'a l'époque actuelle." *Comptes rendus* XXX (1850): 807–812.

Duncan, David. *The Life and Letters of Herbert Spencer.* Reissue. London, 1911.

Elie de Beaumont, Léonce. "Recherches sur quelque-unes des révolutions de la surface du globe." *Ann. des sci. nat.* XVIII (1829): 5–25, 284–416; XIX (1830): 5–99, 177–240.

Falconer, Hugh. "Sivatherium giganteum, a new fossil ruminant genus from the valley of the Markanda, in the Sevalik branch of the sub-Himalayan mountains." *Asiatic Researches, Trans. Asiatic Soc. Bengal* XIX (1836): 1–24.

———. "Note on the fossil hippopotamus of the Sevalik hills." *Asiatic Researches* XIX (1836): 39–53.

———. "On the American fossil elephants of the region bordering on the Gulph of Mexico (*E. columbi.,* Falc.), with general observations on the living and extinct species." *Nat. Hist. Rev.* (1863): 43–114.

Forbes, Edward. Presidential address. *Quart. J. Geol. Soc. Lond.* X (1854): xxii–lxxi.

———. "On the manifestation of polarity in the distribution of organic beings in time." *Notices of the Proceedings of the Royal Institution* I (1851–54): 428–433.

Fourier, J.B.J. "Memoire sur les températures du globe terrestre et des espaces planetaires." *Mém. Acad. Roy des Sci.* VII (1827): 569–604.

Geikie, Sir Archibald. *Life of Sir Roderick I. Murchison.* 2 vols. London, 1875.

———. *Memoir of Sir Andrew Crombie Ramsay.* London, 1895.

Geoffroy Saint Hilaire, Etienne. *Philosophie anatomique: des organes respiratoires sous le rapport de la determination et de l'identité de leur pièces osseuses.* Paris, 1818–22.

———. "Des recherches faites dans les carrières de calcaire oolithiques de Caen, ayant donné lieu à la découverte de plusieurs beaux échantillons et de nouvelles espèces de téléosaurus." *Mém. Acad. Roy. des Sci. XII (1833): 43–61.*

———. "Le degré d'influence du monde ambiant pour modifier les formes animales; question intéressant l'origine des espèces téléosauriennes et successivement celle des animaux de l'époque actuelle." *Mém. Acad. Roy. des Sci.* XII (1833): 63–92.

Geoffroy Saint Hilaire, Isodore. *Vie, traveaux et doctrine scientifique d'Etienne Geoffroy Saint Hilaire.* Paris and Strasburg, 1847.

Grant, Robert E. "Observations on the nature and importance of geology." *Edinb. New Phil. J.* I (1826): 293–302.

———. *Outlines of Comparative Anatomy.* London, 1841.

Haeckel, Ernst. *Naturliche Schopfungs-Geschichte, Gemeinverstendliche wissenschaftliche Vortrage uber die Entwickelungs-Lehre im Allgemeinen und die jenige von Darwin, Goethe und Lamarck im Besonderen.* Berlin, 1873. Translated as

———. *The History of Creation: or the development of the earth and its inhabitants by the action of natural causes. A popular exposition of the doctrine of evolution in general and of that of Darwin, Goethe and Lamarck in particular.* 2 vols. New York, 1876.

———. *Anthropogenie: oder Entwickelungsgeschichte des Menschen . . .* Leipzig, 1874. Translated as

———. *The evolution of Man: A popular exposition of the principal points of human ontogeny and phylogeny.* New York, 1879.

———. *The Last Link. Our present knowledge of the descent of man.* London, 1898.

———. *The Riddle of the Universe at the close of the nineteenth century.* London, 1900.

Harrington, Bernard J. *Life of Sir William E. Logan.* Montreal, 1883.

Herschel, Sir J.F.W. *Preliminary Discourse on the Study of Natural Philosophy.* London, 1830.

D'Holbach, Paul-Henri Thiry, baron. *Système de la nature, ou des lois du monde physique et du monde moral, nouvelle édition avec des notes et des corrections par Diderot.* Paris, 1821. Reprinted with an introduction by Yvon Belavel. Hildesheim: Georg Olms, 1966.

Home, Sir Everard. "Some account of an animal more nearly allied to the fishes than to any other class of animals." *Phil. Trans.* CIV (1814):

571–577. See also CVI (1816): 318–321; CVII (1818): 24–32; and CIX (1819): 209–216.

Hooke, Robert. *The Posthumous Works of Robert Hooke.* London, 1705.

Hooker, Sir Joseph Dalton. "On the origination and distribution of vegetable species—Introductory Essay to the Flora of Tasmania." *Am. J. Sci.* 2nd series, XXIX (1860): 1–25 and 305–326.

Humboldt, Alexander von. *A Geognostical Essay on the Superposition of Rocks.* Translated from the original French. London, 1823.

———. "On petrifactions, or fossil organic remains." *Edinb. Phil. J.* IX (1823): 20–35.

Hunt, Thomas Sterry. "On the mineralogy of certain organic remains from the Laurentian rocks of Canada." *Quart. J. Geol. Soc. Lond.* XXI (1865): 67–71.

Hutton, James. "Theory of the earth; or an investigation of the laws observable in the composition, dissolution and restoration of the land upon the globe." *Trans. Roy. Soc. Edinb.* I, part 2 (1788): 209–304.

———. *Theory of the Earth, with proofs and illustrations.* 1795. Facsimile reprint Weinheim: H. R. Engelmann (J. Cramer) and Codicote, Herts.: Wheldon and Wesley, 1960.

Huxley, Leonard. *Life and Letters of Thomas Henry Huxley.* 3 vols. London, 1908.

———. *Life and Letters of Sir Joseph Dalton Hooker, O.M., G.C.S.I., based on materials collected and arranged by Lady Hooker.* 2 vols. London, 1918.

Huxley, Thomas Henry. "Vestiges of the Natural History of Creation, 10th edition (1853)." *Brit. and For. Med. Chirurg. Rev.* XIII (1854): 332–343.

———. "Geological contemporaneity and persistent types of life." Reprinted from *Quart. J. Geol. Soc. Lond.* (1862) in *Essays,* VIII, pp. 272–304.

———. *Evidences as to Man's Place in Nature.* London, 1863.

———. *Lectures on the Elements of Comparative Anatomy.* London, 1864.

———. "On the animals which are most nearly intermediate between birds and reptiles." Reprinted from *Ann. and Mag. of Nat. Hist.* (1868) in *Scientific Memoirs,* III, pp. 303–313.

———. "Remarks on *Archaeopterix lithographica.*" Reprinted from *Proc. Roy. Soc.* (1868) in *Scientific Memoirs,* III, pp. 340–345.

———. "Palaeontology and evolution." Reprinted from *Quart. J. Geol. Soc. Lond.* (1870) in *Essays,* VIII, pp. 340–388.

———. "On the characters of the pelvis in the mammalia, and the conclusions respecting the origins of the mammals which may be based on them."

Reprinted from *Proc. Roy. Soc.* (1879) in *Scientific Memoirs,* IV, pp. 345–356.

———. *American Addresses, with a lecture on the study of biology.* New York, 1888.

———. "On the reception of the 'Origin of Species.'" In *Life and Letters of Charles Darwin.* II, pp. 179–204.

———. *Collected Essays.* London, 1895.
Vol. II, *Darwiniana.*
Vol. VIII, *Discourses Biological and Geological.*
Vol. IX, *Evolution and Ethics.*

———. *The Scientific Memoirs of Thomas Henry Huxley.* Edited by Sir M. Forster and E. Ray Lankester. 4 vols. London, 1899.

Jameson, Robert. *A System of Mineralogy. Comprehending oryctognosie, geognosie, mineralogical chemistry, mineralogical geography and economical mineralogy.* 3 vols. Edinburgh, 1804–08.

Jones, Thomas Rymer. *General Outline of the Organisation of the Animal Kingdom and manual of comparative anatomy.* 2nd edition. London, 1855.

De Jussieu, Anton Laurent. *Genera Plantarum.* Paris, 1789. Reprinted Weinheim: J. Cramer, 1964.

King, A.T. "Footprints in the coal rocks of Westmoreland Co., Pennsylvania." *Am. J. Sci.* 2nd series, I (1846): 268.

Kirby, William. *On the Power, Wisdom and Goodness of God as manifested in the Creation of Animals and in their History, Habits and Instincts.* 2 vols. London, 1835.

Lamarck, J.B.P.A. de Monet, chevalier de. *Encyclopédie Méthodique Botanique.* 2 vols. Paris and Liege, 1786.

———. *Hydrogeology.* Translated by Albert V. Carozzi. Urbana: University of Illinois Press, 1964.

———. *Histoire naturelle des animaux sans vertèbres.* 4 vols. Paris, 1815–22. Reprinted Brussels: Culture et Civilization, 1969.

———. *Philosophie zoologique, ou exposition des considérations relatives à l'histoire naturelle des animaux* . . . Nouvelle edition révue et precédée d'une introduction biographique par Charles Martins. 2 vols. Paris, 1873.

La Mettrie, Julien Offray de. *Oeuvres philosophiques.* 2 vols. Berlin, 1774.

Linnaeus (Carl von Linné). *Philosophica Botanica.* Stockholm and Amsterdam, 1751. Reprinted by Wheldon and Wesley, Codicote, Herts., and Stechert-Haffner Service Agency, New York, 1966.

———. "On the increase of the habitable earth." In *Select Dissertations from the Amoenetates Academicae, a supplement to Mr. Stillingfleet's tracts relating to natural history.* Translated by F. J. Brand. London, 1781.

———. "Disquisitio de sexu plantarum." *Amoenetates Academicae* X, Erlangen (1790): 100–131

Lock, R.H. *Recent Progress in the Study of Variety, Heredity and Evolution.* London, 1907.

Logan, Sir William E. "On the occurrence of a track and footprints of an animal in the Potsdam sandstone of Lower Canada." *Quart. J. Geol. Soc. Lond.* VII (1851): 247–250.

———. "On the footprints occurring in the Potsdam sandstone of Canada." *Quart. J. Geol. Soc. Lond.* VIII (1852): 199–213.

———. "On the occurrence of organic remains in the Laurentian rock of Upper Canada." *Quart. J. Geol. Soc. Lond.* XXI (1865): 45–50.

Lyell, Sir Charles. "Transactions of the Geological Society of London, volume one, second series." *Quarterly Review* XXXIV (1826): 507–540.

———. *Principles of Geology, being an attempt to explain the former changes of the earth's surface by reference to causes now in operation.* 3 vols. London, 1830–33. 3rd edition. 4 vols. London, 1834.

———. *A Second Visit to the United States of North America.* 2 vols. New York, 1849.

———. Presidential address. *Quart. J. Geol. Soc. Lond.* VII (1851): xxv–lxxvi.

———. *Geological Evidences of the Antiquity of Man, with remarks on theories of the origin of species by variation.* London, 1863.

———. *Principles of Geology, or the modern changes of the earth and its inhabitants considered as illustrative of geology.* 2 vols. London, 1867–8.

———. *Sir Charles Lyell's Journals on the Species Question.* Edited by Leonard G. Wilson. New Haven and London: Yale University Press, 1970.

de Maillet, Benôit. *Telliamed or conversations between an Indian philosopher and a French missionary on the diminution of the sea . . .* Translated and edited by Albert V. Carrozi. Illinois University Press, 1968.

Mantell, Gideon Algernon. "Notice on the *Iguanodon,* a newly discovered fossil reptile from the sandstone of Tilgate forest in Sussex." *Phil. Trans.* CXXV (1825): 179–186.

———. "The geological age of reptiles." *Edinb. New Phil. J.* XI (1831): 181–185.

———. *The Wonders of Geology, or a familiar exposition of geological phenomena.* 3rd edition. 2 vols. London, 1839.

———. "Description of the *Telerpeton elginense,* a fossil reptile recently discovered in the Old Red sandstone of Moray . . ." *Quart. J. Geol. Soc. Lond.* VIII (1852): 100–109.

Marsh, Othniel C. "Preliminary description of *Hesperornis regalis . . ." Am. J. Sci.* 3rd series, III, whole number CIII (1872): 360–365. See also IV, whole number CIV (1872): 344, and V, whole number CV (1873): 74.

———. "On a new sub-class of fossil birds (Odontornithes)." *Am. J. Sci.* V (1873): 161–162.

———. "Notice of some new equine mammals from the Tertiary formation." *Am. J. Sci.* VII (1874): 247–258.

———. "On the Odontornithes, or birds with teeth." *Am. J. Sci.* X (1875): 403–408.

———. "Notice of some new Tertiary mammals." *Am. J. Sci.* XII (1876): 401–404.

———. "Recent discoveries of extinct mammals." *Am. J. Sci.* XII (1876): 59–61.

———. *Introduction and Succession of Vertebrate Life in America. An address delivered before the American Association for the Advancement of Science at Nashville, Tenn., August 30th 1877.* No place or date of publication.

———. *Odontornithes: a monograph on the extinct toothed birds of North America.* Report of the Geological Exploration of the Fortieth Parallel, vol. VII. Washington, 1880.

———. *Dinocerata. A monograph of an extinct order of gigantic mammals.* Monographs of the U.S. Geological Survey, vol. X. Washington, 1886.

Maupertuis, P.L. Moreau de. *Oeuvres de M. de Maupertuis.* Dresden, 1752.

———. *Oeuvres.* 4 vols. Lyons, 1768. Reprinted by Georg Holms, Hildesheim, 1965.

Meyer, Hermann von. "The reptiles of the coal formation." *Quart. J. Geol. Soc. Lond.* IV (1848) part 2: 51–56.

———. "On the *Archegosaurus* of the coal formation." *Quart. J. Geol. Soc. Lond.* VI (1850) part 2: 58–59.

Miller, Hugh. *The Old Red Sandstone, or new walks in an old field.* Edinburgh, 1841. New and enlarged edition. Boston, 1858.

———. *Footprints of the Creator, or the Asterolepis of Stromness.* 3rd edition. London, 1850.

———. *The Testimony of the Rocks, or geology in its bearings on the two theologies, natural and revealed.* 38th thousand. Edinburgh, 1870.

Mivart, St. George Jackson. *The Genesis of Species.* 2nd edition. London, 1871.

———. "On the possible dual origin of the Mammalia." *Proc. Roy. Soc.* XLIII (1887–88): 372–379.

Murchison, Sir Roderick I. *The Silurian System, founded on geological researches in the counties of Salop, Hereford, Radnor, Montgomery,*

Caermarthen, Brecon, Pembroke, Monmouth, Gloucester, Worcester and Stafford; with descriptions of the coalfields and overlying formations. London, 1839.

———. *Siluria. The history of the oldest known rocks containing organic remains, with a brief sketch of the distribution of gold over the earth.* London, 1854. 5th edition. London, 1872.

Oken, Lorenz. *Elements of Physicophilosophy.* London, 1847.

Owen, Richard. "Observations on the fossils representing *Thylacotherium prevosti* (Valenciennes) with reference to the doubts of its mammalian and marsupial nature recently promulgated and on *Phascolotherium bucklandi." Trans. Geol. Soc. Lond.* VI (1842): 47–66. See also *Proc. Geol. Soc. Lond.* III (1838): 5–9, 17–23, and 61–98.

———. "Report on British fossil reptiles, part 2." In *Report . . . of the British Association for the Advancement of Science . . . 1841.* Pp. 60–204. London, 1842.

———. *A History of British Fossil Mammals and Birds.* London, 1846.

———. *Lectures on the Comparative Anatomy and Physiology of the Vertebrate Animals, part 1, Fishes.* London, 1846.

———. *On the Archetype and Homologies of the Vertebrate Skeleton.* London, 1848.

———. *On the Nature of Limbs. A discourse delivered on Friday, February 9th at an evening meeting of the Royal Institution of Great Britain.* London, 1849.

———. "Lyell on life and its successive development." *Quarterly Review* LXXXIX (1851): 412–451.

———. Letter "On the track of a quadruped imprinted on lower Silurian sandstone, from Beauharnois, 20 miles above Montreal (Chelonian)." *Quart. J. Geol. Soc. Lond.* VII (1851): lxxv–lxxvi.

———. "Description of the impressions in the Potsdam sandstone discovered by Mr. Logan in Lower Canada." *Quart. J. Geol. Soc. Lond.* VII (1851): 250–252.

———. "Descriptions of the impressions and footprints of the Protichnites from the Potsdam sandstone of Canada." *Quart. J. Geol. Soc. Lond.* VIII (1852): 214–225.

———. "On some fossil reptilian and mammalian remains from the Purbecks." *Quart. J. Geol. Soc. Lond.* X (1854): 420–433.

———. "Description of the skull of *Dicynodon* (*D. tigriceps*, Ow.), transmitted from South Africa by A.G. Bain, Esq.," *Trans. Geol. Soc. Lond.* ser. 2, VII (1845–56): 233–240.

———. "Darwin on the Origin of Species." *Edinb. Review* CXI (1860): 487–532.

———. *Palaeontology, or a systematic study of extinct animals and their geological relations.* Edinburgh, 1860.

———. "On the *Archaeopterix* of von Meyer . . ." *Phil. Trans.* CLIII (1864): 33–47.

———. *On the Anatomy of the Vertebrates.* 3 vols. London, 1866–68.

Owen, Rev. Richard. *The Life of Richard Owen.* 2 vols. London, 1894.

Paley, William. *Natural Theology; or evidences of the existence and attributes of the Deity collected from the appearances of nature.* London, 1803.

Parkinson, James. *Organic Remains of a former world, or examination of the mineralised remains of the vegetation and animals of the antediluvian world.* I, 2nd edition. London, 1820. II and III. London, 1808–11.

Phillips, John. *Figures and Descriptions of the Palaeozoic Fossils of Cornwall, Devon and West Somerset, observed in the course of the Ordnance Geological Survey of that district.* London, 1841.

———. *Memoir of William Smith.* London, 1844.

———. *Life on the Earth, its origin and succession.* London and Cambridge, 1860.

Pictet, Françoise Jules. *Traité élémentaire de paléontologie, ou histoire naturelle des animaux fossiles considerés dans leurs rapports zoologiques et géologiques.* 3 vols. Paris, 1844.

———. "Sur l'origine de l'espèce par Charles Darwin." *Archives des sciences physiques et naturelles, Bibliotheque Universelle de Genève* nouvelle période, VII (1860): 233–255.

Playfair, John. *Illustrations of the Huttonian Theory of the Earth.* Edinburgh, 1802. Reprinted New York: Dover, 1964.

Powell, Baden. *Essays on the Spirit of the Inductive Philosophy, the unity of worlds and the philosophy of creation.* London, 1855.

Prévost, Constant. "Observations sur les schistes oolithiques de Stonesfield en Angleterre dans lesquelles ont trouvés plusieurs ossemens fossiles de mammifères." *Ann. des sci. nat.* V (1825): 389–417.

———. "De la chronologie des terrains et du synchronisme des formations." *Comptes rendus* XX (1845): 1062–1071.

———. "Quelques propositions relatives à l'état originaire et actuel de la masse terrestre, à la formation du sol, aux causes qui ont modifié le relief de sa surface, aux êtres qui l'ont successivement habité." *Comptes rendus* XXXI (1850): 461–469.

Richardson, G.F. *An Introduction to Geology and its associate sciences, mineralogy, fossil botany and palaeontology.* London, 1851.

Robinet, J.B. *De la nature.* Vol. I. Nouvelle édition. Amsterdam, 1763. Vols. II, III, and IV. Amsterdam, 1766.

Schelling, F.W.J. *The Ages of the World.* Translated by F. Bolman, Jr. New York, 1942

Sedgwick, Adam. Presidential address. *Proc. Geol. Soc. Lond.* I (1830): 182–212.

———. Presidential address. *Proc. Geol. Soc. Lond.* I (1831): 281–316.

———. *Syllabus of a Course of Lectures on Geology.* 3rd edition. Cambridge, 1837.

———. "Synopsis of the English series of stratified rocks inferior to the old Red Sandstone." *Proc. Geol. Soc. Lond.* II (1838): 675–685.

———. "Vestiges of the Natural History of Creation." *Edinb. Review* LXXXII (1845): 1–85.

———. *A Discourse on the studies of the University of Cambridge.* 5th edition, with additions and a preliminary dissertation. London and Cambridge, 1851.

———. Letter in *The Spectator* XXXII (24 March 1860): 285.

Sedgwick, A. and Murchison, Roderick I. "On the structure and relations of the deposits contained between the primary rocks and the Oolitic series in the north of Scotland." *Trans. Geol. Soc. Lond.* 2nd series, III (1835): 125–160.

Seeley, Harry Govier. "On the classification of the fossil animals commonly named Dinosauria." *Proc. Roy. Soc.* XLII (1887–88): 165–171.

———. "Researches on the structure, organization and classification of the fossil Reptilia—II on *Pareiasaurus bombidens* (Owen), and the significance of its affinities to amphibians, reptiles and mammals." *Phil. Trans.* CLXXIX B (1888): 59–109.

———. "Researches . . .—III on parts of the skeleton of a mammal from the Triassic rocks of Klipfontein, Fraserberg, S. Africa (*Theriodesmus phylarchus,* Seeley) illustrating the reptilian inheritance of the mammalian hand." *Phil. Trans.* CLXXIX B (1888): 141–155.

———. "Researches . . .—IX section 2, the reputed mammals from the Karoo formations of the Cape Colony." *Phil. Trans.* CLXXXV B (1894): 1019–1028.

Smiles, Samuel. *Robert Dick, baker, of Thurso, geologist and botanist.* New York, 1879.

Smith, William. *A memoir to the map and delineation of the strata of England & Wales, with a part of Scotland.* London, 1815.

———. *Strata Identified by Organised Fossils.* London, 1816.

———. *Stratigraphical System of Organised Fossils, with reference to the specimens of the original geological collection of the British Museum.* London, 1817.

Spencer, Herbert. "The development hypothesis." Reprinted from *The Leader* (1852) in *Illustrations of Universal Progress.* pp. 377–383.

―――. "Progress: its law and cause." Reprinted from the *Westminster Review* (1857) in *Illustrations of Universal Progress. pp. 1–60.*

―――. "Illogical geology." Reprinted from the *Universal Review* (1859) in *Illustrations of Universal Progress.* pp. 325–376.

―――. *First Principles of a New Philosophy.* New York, 1864.

―――. *Principles of Biology.* 2 vols. London, 1864.

―――. *Illustrations of Universal Progress, a series of discussions.* New York, 1864.

―――. *An Autobiography.* 2 vols. New York, 1904.

Valenciennes, Achille. "Observations sur les mâchoires fossiles des couches Oolithiques de Stonesfield nommés *Didelphis prevosti* et *Didephis bucklandi."* *Comptes rendus* VII (1838): 572–580.

Voltaire, Francois Marie Arouet. *Oeuvres complètes.* 13 vols. Paris, 1876.

Wallace, Alfred Russel. "On the law which has regulated the introduction of new species." Reprinted from *Ann. and Mag. of Nat. Hist.* (1855) in *Natural Selection and Tropical Nature,* pp. 3–19.

―――. "On the tendency of varieties to depart indefinitely from the original type." Reprinted from *J. Linn. Soc.* (1858) in *Natural Selection and Tropical Nature,* pp. 20–33.

―――. "Sir Charles Lyell on geological climates and the origin of species." *Quarterly Review* (American edition) CXXVI (1869): 187–205.

―――. "The limits of natural selection applied to man." Reprinted from *Contributions to the Theory of Natural Selection* (1870) in *Natural Selection and Tropical Nature,* pp. 186–214.

―――. *Darwinism, an exposition of the theory of Natural Selection with some of its applications.* London, 1890.

―――. *Natural Selection and Tropical Nature.* New edition. London, 1895.

Whewell, William. "Lyell's *Principles of Geology,* Vol. I." *British Critic* IX (1831): 180–206.

―――. "Lyell's *Principles of Geology,* Vol. II." *Quarterly Review.* XLVII (1832): 103–132.

―――. *Astronomy and General Physics considered with reference to Natural Theology.* 3rd edition. London, 1834.

―――. *History of the Inductive Sciences.* 3 vols. London, 1837.

―――. Presidential address. *Proc. Geol. Soc. Lond.* III (1838): 61–98.

―――. *Philosophy of the Inductive Sciences.* 2nd edition. London, 1847.

Whiston, William. *A New Theory of the earth, from its Original to the consummation of all things.* London, 1696.

Wilberforce, Samuel, Bishop of Oxford. "Darwin's Origin of Species." *Quarterly Review* CVIII (1860): 225–264.

Woodward, A.S. and Sherborn, C.D. *A Catalogue of British Fossil Vertebrata.* London, 1890.

Secondary Sources

Adams, Frank Dawson. *The Birth and Development of the Geological Sciences.* Reprinted New York: Dover, 1954.

Appleman, Philip, (ed.). *Darwin, a Norton critical edition.* New York: Norton, 1970.

Arkell, William J. *The Jurassic System in Great Britain.* Oxford: Clarendon Press, 1933.

Barnett, S.A., (ed.). *A Century of Darwin.* Reprinted London: Mercury Books, 1962.

Bartholomew, Michael. "Lyell and evolution: an account of Lyell's response to the prospect of an evolutionary ancestry for man." *Brit. J. Hist. Sci.* VI (1973): 261–303.

de Beer, Sir Gavin. "Darwin and embryology." In Barnett (ed.), *A Century of Darwin.* Pp. 153–172.

——. *Charles Darwin.* London: Nelson, 1963.

Bell, P.R., (ed.). *Darwin's Biological Work.* Reprinted New York: Wiley, 1964.

Birembaut, Arthur. "Elie de Beaumont." In Gillispie (ed.), *Dictionary of Scientific Biography.* II, pp. 347–350.

Bourdier, Franck. "Geoffroy Saint-Hilaire versus Cuvier; the campaign for paleontological evolution (1825–1838)." In Schneer (ed.), *Toward a History of Geology.* Pp. 36–61.

Bowler, Peter J. "Buffon and Bonnet: theories of generation and the problem of species." *J. Hist. Biology* VI (1973): 259–281.

——. "Evolutionism in the Enlightenment." *History of Science* XII (1974): 159–183.

——. "The changing meaning of 'evolution.'" *J. Hist. Ideas* XXXVI (1975): 95–114.

Brunet, Pierre. *Maupertuis.* Paris: Albert Blanchard, 1929.

Burkhardt, R.W. Jr. "Lamarck, evolution and the politics of science." *J. Hist. Biol.* III (1970): 275–298.

——. "The inspiration of Lamarck's belief in evolution." *J. Hist. Biol.* V. (1972): 413–438.

Bury, J.B. *The Idea of Progress, an inquiry into its growth and origin.* Reprinted New York: Dover, 1955.

Cahn, Theophile. *La vie et l'oeuvre d'Etienne Geoffroy Saint-Hilaire*. Paris: Presses Universitaires de France, 1962.

Cannon, Walter F. "The problem of miracles in the 1830s." *Victorian Studies* IV (1960): 5–32.

———. "The uniformitarian-catastrophist debate." *Isis* LI (1960): 38–55.

———. "The impact of uniformitarianism. Two letters from John Herschel to Charles Lyell, 1836–1837." *Proc. Am. Phil. Soc.* CV (1961): 301–314.

———. "William Buckland." In Gillispie (ed.), *Dictionary of Scientific Biography*. II, pp. 566–572.

Canguilhem, Georges, *et al.* "Du developpement a l'evolution au XIXe siècle." *Thales* II (1960): 3–68.

Colbert, E. H. *Men and Dinosaurs. The search in field and laboratory.* Reprinted Harmondsworth: Penguin Books, 1971.

Coleman, William. "Lyell and the reality of species." *Isis* LIII (1962): 325–338.

———. *Georges Cuvier, Zoologist. A study in the history of evolution theory.* Cambridge, Mass: Harvard University Press, 1964.

———. *Biology in the Nineteenth Century. Problems of Form, Function and Transformation.* New York, London, Sydney, Toronto: Wiley History of Science series, 1971.

Conry, Y. *Correspondence entre Charles Darwin et Gaston de Saporta. Précédé d'une histoire de la paléobotanique en France en XIXe siècle.* Paris: Presses Universitaires de France, 1972.

Coppleston, Fr. F. *A History of Philosophy.* VII, part I. New York: Image Books, 1965.

Crocker, Lester G. "Diderot and eighteenth century French transformism." In Glass (ed.), *Forerunners of Darwin.* Pp. 114–143.

Daudin, Henri. *Etudes d'histoire des sciences naturelles.* Vol. I. *De Linné a Jussieu, méthodes de la classification et l'idée de série en botanique et en zoologie.* Vol. II. *Cuvier et Lamarck, les classes zoologiques et l'idée de série animale.* Paris, 1926.

Delair, Justin B. and Sargeant, William A.S. "The earliest discoveries of dinosaurs." *Isis* LXVI (1975): 5–25.

Dupree, A. Hunter. *Asa Gray.* Cambridge, Mass.: Harvard University Press, 1959.

Edwards, W.N. *The Early History of Palaeontology.* London: British Museum (Natural History), 1967.

Egerton, Frank N. "Refutation and conjecture: Darwin's response to Sedgwick's attack on Chambers." *Studies in History and Philosophy of Science* I (1970): 176–183.

Eiseley, Loren. *Darwin's Century. Evolution and the men who discovered it.* New York: Doubleday, 1958.

Ellegård, Alvar. *Darwin and the General Reader. The reception of Darwin's theory of evolution in the British periodical press. 1859–1872.* Göteburg, 1958.

Eyles, Joan M. "William Smith: some aspects of his life and work." In Schneer (ed.), *Toward a History of Geology.* Pp. 142–158.

Farber, Paul L. "Buffon and the concept of species." *J. Hist. Biol.* V (1972): 259–284.

Foucault, Michel. *The Order of Things. An archaeology of the human sciences.* New York: Pantheon Books, 1972.

Geikie, Sir Archibald. *The Founders of Geology.* London, 1897. Reprinted New York: Dover, 1962.

Gerstner, Patsy A. "Vertebrate paleontology, an early nineteenth century transatlantic science." *J. Hist. Biol.* III (1970): 137–148.

Gillispie, C.C. "The formation of Lamarck's evolutionary theory." *Arch. Int. d'hist. des sci.* IX (1956): 323–338.

———. (ed.). *Dictionary of Scientific Biography.* 9 vols. issued New York: Scribner, 1970–75.

———. *Genesis and Geology, a study in the relations of scientific thought, natural theology, and social opinion in Great Britain, 1790–1850.* Reprinted New York: Harper, 1959.

Glass, Bentley (ed.). *Forerunners of Darwin 1745–1859.* Baltimore: The Johns Hopkins Press, 1959. Reprinted 1968.

———. "Heredity and variation in the eighteenth century concept of the species." In Glass (ed.), *Forerunners of Darwin.* Pp. 144–172.

Goudge, T.A. *The Ascent of Life. A philosophical study of the theory of evolution.* Toronto: University of Toronto Press, 1961.

Gould, Stephen Jay. "Alpheus Hyatt." In Gillispie (ed.), *Dictionary of Scientific Biography.* VII, pp. 613–614.

Greene, John C. *The Death of Adam. Evolution and its impact on Western Thought.* Iowa City: University of Iowa Press, 1959.

———. "The Kuhnian paradigm and the Darwinian revolution in natural history." in Roller (ed.), *Perspectives in the History of Science.* Pp. 3–25.

Gruber, Jacob. *A Conscience in Conflict. The life of St. George Jackson Mivart.* New York: Columbia University Press, 1960.

Haber, F.C. "Fossils and the idea of a process of time in natural history." In Glass (ed.), *Forerunners of Darwin.* Pp. 222–264.

———. *The Age of the World, Moses to Darwin.* Baltimore: Johns Hopkins Press, 1959.

Hampson, Norman. *The Enlightenment.* Harmondsworth: Penguin Books, 1968.

Hanson, Bert. "Bronn, Heinrich Georg." In Gillispie (ed.), *Dictionary of Scientific Biography.* II, pp. 497–498.

Harrison, James. "Erasmus Darwin's view of evolution." *J. Hist. Ideas* XXXII (1972): 247–264.

Heim, R. (ed.). *Buffon.* Paris: Muséum d'Histoire naturelle, 1952.

Himmelfarb, Gertrude. *Darwin and the Darwinian Revolution.* London: Chatto & Windus, 1959.

Hodge, M.J.S. "Lamarck's science of living bodies." *Brit. J. Hist. Sci.* V (1971): 323–352.

——. "The universal gestation of nature: Chambers' *Vestiges* and *Explanations.*" *J. Hist. Biol.* V (1972): 127–152.

Hooykaas, R. "The parallel between the history of the earth and the history of the animal world." *Arch. int. d'hist. des sci.* X (1957): 3–18.

——. *Natural Law and Divine Miracle. The principle of uniformity in geology, biology and theology.* Leiden: Brill, 1959.

——. "Geological uniformitarianism and evolution." *Arch. int. d'hist. des sci.* XIX (1966): 3–19.

Hull, D. L. *Darwin and his Critics. The reception of Darwin's theory of evolution by the scientific community.* Cambridge, Mass.: Harvard University Press, 1973.

Huxley, Julian S. *Evolution in Action, based on the Patten Foundation lectures delivered at Indiana University in 1951.* London: Chatto and Windus, 1953.

——. *Evolution: the Modern Synthesis.* 2nd edition. London: Allen & Unwin, 1963.

Irvine, W. *Apes, Angels and Victorians: a joint biography of Darwin and Huxley.* Reprinted Cleveland: Meridian, 1959.

Keith, Sir Arthur. *Darwin Revalued.* London: Watts & Co., 1955.

Kermack, D.M. and Kermack, K.A. (eds.). *Early Mammals.* Supplement 1 to *Zool. J. Linn. Soc.* L (1971).

King-Hele, Desmond. *Erasmus Darwin.* New York: Scribner, 1963.

Leroy, Jean Francois. "Brongniart, Adolphe Theodore." In Gillispie (ed.), *Dictionary of Scientific Biography.* II, pp. 491–493.

Lovejoy, A.O. *The Great Chain of Being. A study in the history of an idea.* Reprinted New York: Harper, 1960

——. "Buffon and the problem of species." In Glass (ed.), *Forerunners of Darwin.* Pp. 84–113.

——. "The argument for organic evolution before the *Origin of Species,* 1830–1858." in Glass (ed.), *Forerunners of Darwin.* Pp. 356–414.

————. "Schopenhauer as an evolutionist." in Glass (ed.) *Forerunners of Darwin.* Pp. 415–437.

————. "Recent criticism of the Darwinian theory of recapitulation—its grounds and its initiator." in Glass (ed.), *Forerunners of Darwin.* Pp. 438–458.

Lurie, Edward. *Louis Agassiz. A life in science.* Chicago: University of Chicago Press, 1960.

————. "Louis Agassiz and the idea of evolution." *Victorian Studies* III (1959–60): 87–108.

MacLeod, Roy M. "Evolutionism and Richard Owen." *Isis* LVI (1965): 259–280.

Mandelbaum, Maurice. *History, Man, and Reason, a study in nineteenth century thought.* Baltimore and London: The Johns Hopkins Press, 1971.

Mayr, Ernst. "Agassiz, Darwin and evolution." *Harvard Library Bulletin* XIII (1959): 165–194.

————. "Lamarck revisited." *J. Hist. Biology* V (1972): 55–94.

McKinney, H. Lewis. *Wallace and Natural Selection.* New Haven and London: Yale University Press, 1972.

McPherson, Thomas. *The Argument from Design.* London: Macmillan, 1972.

Millhauser, Milton. *Just before Darwin, Robert Chambers and Vestiges.* Middletown, Conn.: Wesleyan University Press, 1959.

Nieuwenkamp, W. "Leopold von Buch." In Gillispie (ed.), *Dictionary of Scientific Biography.* II, pp. 552–557.

North F.J. *Sir Charles Lyell, Interpreter of the principles of geology.* London: Arthur Barker Ltd., 1965.

O'Brien, Charles F. "Sir William Dawson. A life in science and religion." *Am. Phil. Soc. Memoirs.* LXXXIV 1971.

————. "*Eozoön Canadense,* the dawn animal of Canada." *Isis* LXI (1971): 200–223.

Oppenheimer, Jane. "An embryological enigma in the *Origin of Species.*" In Glass (ed.), *Forerunners of Darwin.* Pp. 292–322.

Osborne, Henry Fairfield. *Cope: Master Naturalist. The life and letters of Edward Drinker Cope, with a bibliography of his writings classified by subject.* Princeton: Princeton University Press, 1931.

Peel, J.B.Y. *Herbert Spencer, the evolution of a sociologist.* London: Heinemann, 1971.

Piveteau, Jean. "Le débat entre Cuvier et Geoffroy Saint-Hilaire sur l'unité de plan et de composition." *Revue d'hist. des sci.* III (1950): 346–363.

Pollard, Sydney. *The Idea of Progress, History and Society.* Reprinted Harmondsworth: Penguin Books, 1971.

Raikov, Boris E. *Karl Ernst von Baer, 1792–1876, sein Leben and sein Werk.* Acta Historia Leopoldina No. 5. Leipzig: J.A. Barth, 1968.

Roberts, F.J. *Plant Hybridization before Mendel.* Princeton: Princeton University Press, 1929.

Roger, Jacques. (ed.). Buffon, *Les Epoques de la nature, Mémoires du Muséum National d'Histoire Naturelle.* Nouvelle série, serié C, X. Paris, 1962.

———. *Les sciences de la vie dans la penseé francaise du XVIII^{e}siècle.* Paris: Armand Colin, 1963. 2nd edition. 1972

Roller, Duane H.D. (ed.). *Perspectives in the History of Science & Technology.* Norman: University of Oklahoma Press, 1971.

Romer, A.F. "Darwin and the fossil record." In Barnett (ed.), *A Century of Darwin.* Pp. 130–152.

Rudwick, M.J.S. "A critique of uniformitarian geology: a letter from W.D. Conybeare to Charles Lyell, 1841." *Proc. Am. Phil. Soc.* CXI (1967): 272–287.

———. Essay review of Louis Agassiz, *Studies on Glaciers,* translated by Albert V. Carozzi. *History of Science* VIII (1969): 136–157.

———. "The strategy of Lyell's *Principles of Geology.*" *Isis* LXI (1970): 5–33.

———. "Brongniart, Alexandre." In Gillispie (ed.), *Dictionary of Scientific Biography.* II, pp. 493–497.

———. "Uniformity and progression: reflections on the structure of geological theory in the age of Lyell." In Roller (ed.), *Perspectives in the History of Science & Technology.* Pp. 209–227.

———. *The Meaning of Fossils. Episodes in the history of palaeontology.* London: MacDonald and New York: American Elsevier, 1972.

Russell, E.S. *Form and Function. A contribution to the history of animal morphology.* London: John Murray, 1916.

Sanford, William F., Jr. "Dana and Darwinism." *J. Hist. Ideas* XXVI (1965): 531–546.

Savioz, Raymond. *La philosophie de Charles Bonnet de Genève.* Paris: J. Vrin, 1948.

Schiller, J. (ed.). *Colloque international "Lamarck," tenue au Muséum national d'Histoire naturelle.* Paris: Blanchard, 1971.

———. "L'échelle des etres et la série chez Lamarck." In *Colloque international "Lamarck,"* pp. 87–103.

Schneer, Cecil J. (ed.). *Toward a History of Geology.* Cambridge, Mass: M.I.T. Press, 1969.

Schuchert, Charles, and LeVene, Clara Mae. *O.C. Marsh, pioneer in paleontology*. New Haven: Yale University Press, 1940.

Sherwood, Morgan B. "The Dana-Lewis controversy, 1856–1857, Genesis, evolution and geology." In Schneer (ed.), *Toward a History of Geology*. Pp. 305–316.

Simpson, G.G. *A Catalogue of the Mesozoic Mammals*. London: British Museum (Natural History), 1928.

———. *The Meaning of Evolution. A study in the history of life and its significance for man*. New Haven and Oxford: Yale and Oxford University Presses, 1949.

———. *This View of Life. The world of an evolutionist*. Reprinted New York: Harbinger, 1964.

———. "Mesozoic mammals revisited." In Kermack (ed.), *Early Mammals*, pp. 181–198.

Swinton, W.E. "Harry Govier Seeley and the Karoo reptiles." *Bull. Brit. Mus. (Nat. Hist.) Historical Series* III (1962), No. 1: 1–39.

———. *The Dinosaurs*. 2nd edition, London: Allen and Unwin, 1970.

Teilhard de Chardin, Pierre. *The Phenomenon of Man*. London: Collins, 1959.
Temkin, Owsei. "German concepts of ontogeny and history arouna 18vv. *Bull. Hist. Med.* XXIV (1950): 227–246.

———. "The idea of descent in post-romantic German biology." In Glass (ed.), *Forerunners of Darwin*. Pp. 323–355.

Thoday, J.M. "Natural selection and biological progress." In Barnett (ed.), *A Century of Darwin*. Pp. 313–333.

White, Andrew Dickson. *A History of the Warfare of Science with Theology in Christendom*. 2 vols. 1896. Reprinted New York: Dover, 1960.

Wilkie, J.S. "The idea of evolution in the writings of Buffon." *Annals of Science* XII (1956): 48–62, 212–227, and 255–266.

Wilson, L.G. "The origins of Charles Lyell's uniformitarianism." *Geol. Soc. of America Special Paper* LXXXIX (1967): 35–62.

———. "The intellectual background in Charles Lyell's *Principles of Geology*. 1830–1833." In Schneer (ed.), *Toward a History of Geology*. Pp. 426–443.

———. (ed.). *Sir Charles Lyell's Journals on the Species Question*. New Haven and London: Yale University Press, 1970.

———. "Sir Charles Lyell on the species question." *American Scientist* LIX (1971): 43–55.

———. *Charles Lyell: The Years to 1841: The revolution in geology*. New Haven: Yale University Press, 1972.

Young, R.M. "Darwin's metaphor: does nature select?" *The Monist* LV (1971): 442–503.

Zittel, Karl von. *History of Geology and Palaeontology*. Reprinted Weinheim: J. Cramer, 1962.

Index of Names